ACTS OF SUPREMACY

⚓

A CAUTIONARY INDICTMENT OF AMERICA'S MILITARY JUSTICE SYSTEM

WALTER FRANCIS FITZPATRICK, III

Edited by Dominic McFarland Martin

DEFIANCE PRESS
& PUBLISHING

DEFIANCE PRESS
& PUBLISHING

ISBN-13: 978-1-959677-70-3 (Paperback)
ISBN-13: 978-1-959677-71-0 (Hardcover)
ISBN-13: 978-1-959677-69-7 (eBook)

Published by Defiance Press and Publishing, LLC

Bulk orders of this book may be obtained by contacting Defiance Press and Publishing, LLC. www.defiancepress.com.

Public Relations Dept. – Defiance Press & Publishing, LLC
281-581-9300
pr@defiancepress.com

Defiance Press & Publishing, LLC
281-581-9300
info@defiancepress.com

DEDICATIONS

I have written this tale first and foremost for my father, Walter Francis Fitzpatrick Jr., MC, USNR, retired, and for Michael Brent Nordeen and his brother William, navy aviators and captains both.

Also for the officers and men of the *USS Indianapolis* (CA-35).

And for all men afloat.

And the sea shall know your names.

May there always be a *USS Indianapolis* (CA-35)!

"The remedy for anger is patience."

—Geoffrey Chaucer (1343–1400),
The Canterbury Tales (1387–1400)

⚓

"Beware the fury of a patient man!"

—John Dryden (1631–1700),
Absalom and Achitophel, 1 (1681)

TABLE OF CONTENTS

PROLOGUE

I, WALTER FRANCIS FITZPATRICK III, do solemnly swear that I will support and defend the Constitution of the United States against all enemies, foreign and domestic; that I will bear true faith and allegiance to the same; that I take this obligation freely, without any mental reservation or purpose of evasion; and that I will well and faithfully discharge the duties of the office upon which I am about to enter; So help me God.

—The Oath of Office for Officers, affirmed to defend the Constitution, not a singular man or individual office holder.

TO ALL VETERANS

I have eaten your bread and salt.

I have drunk your water and wine.

The deaths I watched beside.

And the lives ye lived were mind.

Was there aught that I did not share

In vigil or toil or ease, —

One joy or woe that I did not know,

Dear hearts across the seas?

I have written the tale of our life

For a sheltered people's mirth

In jesting guise—but ye are wise,

And ye know what the jest is worth.

—Rudyard Kipling (1865–1936),[1] Prelude, Stanza 1.
Departmental Ditties and Other Verses (1836)

---

1 Kipling, Rudyard. (1892, 1893, 1899, 1917). Departmental Ditties and Ballads and Bar-rack Room Ballads. New York: McMillan And Co, "PRELUDE."

⚓

The most sacred thing I will do in my life is honor my father, Medical Doctor and Navy Captain: Walter Francis Fitzpatrick Junior.

⚓

Laying myself before open inquiry, anticipating questions respecting the timing of this work, stated simply, this publication has had to find within me its proper place in time and context. Yet, I am a patient man. Dear reader, please know this: Finally, the time has come for a full explication on the subject of our unfair, unregulated system of military discipline.

⚓

For centuries, a select group of officers has worked frantically and furiously to maintain a cloak of secrecy around the sensitive inner workings of our military discipline system. Therefore, today, the average citizen generally has scant idea of those closeted, complicated, and clandestine machinations. Until now, our citizenry has had absolutely no idea as to how the military hierarchy works to fix court-martials to convict men, many of whom may have been unjustly charged.

Thus, this work finds its place in exposing and eliminating what is now termed the "deep state." To be more precise, this book is written to expose and eliminate the military components of the deep state.

To step back in time, the existence of the United States Naval Academy at Annapolis is the result of a covert Act of Supremacy.

In 1970, I was simply another fresh-faced navy-enlisted man just starting my naval career, enrolled in boot camp at the Great Lakes Naval Training Center in Lake County, Illinois. One point driven home hard and deeply early on during that initial training phase was that I and all other military servicemen had signed on to defend the United States Constitution, but that, ironically, none of us were fully protected by its words. Those in the military, we were then instructed, are a special class of people, one set apart entirely from the civilian community that we had just left—and, instead, we were bound to a separate set of time-honored Acts of Supremacy[2], sets of various strictures, rules, and regulations. During boot camp in that long-ago summer of 1970, Machinist Mate Chiefs MacDonald and Osborne drilled into the men of Company 999 that it was our job to defend the US Constitution with our lives despite the fact that, at the same time, those same lives were not protected by that key document. Neither MacDonald nor Osborne explained to us just how this special, quixotic circumstance had come to pass. Contradictorily, our two drill instructors were most enthusiastic in drilling into their charges the true meaning and practice of blind, unquestioning obedience.

That same military maxim was repeated to the 1,360 or so plebes (Nota bene: "Plebes" are newly arrived cadets. A plebe, short for "plebian," is a person who is considered quite ordinary. The word is pronounced "plebe.") who entered the Naval Academy of Annapolis, Maryland, in the Class of 1975, also without a good explanation. In terms of rank or authority, not one of those plebes was then in a position to ask salient questions as to why or how this lower-class status of sailors had been first engendered and later institutionalized.

During any plebe year, the young first-year students must memorize huge volumes of material; thus, their attention is diverted away from the reality that midshipmen are no longer free men as the Constitution had intended. Indeed, today I still hold my issued copy of *Reef Points*, the veritable Bible issued to all plebes at the United States Naval Academy when they first arrive on campus. Throughout their first year at the academy, plebes are aggressively quizzed on how well they have been able to be memorize nearly everything in this miniaturized encyclopedia, one

---

2 Penned by Henry VIII.

containing history, jargon, weapons inventory, facts, figures, details, and trivia tied to the Navy and Marine Corps, with a special emphasis on the Naval Academy itself.

The gap between what the Constitution promises and what the academy delivered only grew wider as the four-year matriculation proceeded.

Today, I do recall the pertinent words of Captain Frank Ramsey:

"We're here to preserve democracy, not practice it."

—*Crimson Tide* (book and movie), Story by Richard Henrick (1995), Screenplay by Michael Schiffer (1995)

⚓

After graduation in 1975 (and paraphrasing Psalm 107: 23–31), I went down to the sea in ships as a surface warfare officer. For me, this was a new training experience with ever-increasing challenges, and therefore I possessed neither the time nor the place to study why the military discipline system did not reconcile or mesh well with our Constitution. After that summer of 1975, no questions regarding that unquestioned and unflinching obedience was allowed by senior officers, and that training experience was quite common amongst all the sailors with whom I served.

However, some fifteen years later, experiencing Rear Admiral John Bitoff's betrayal as he court-martialed me was a life-changing experience. One of Bitoff's unintended results was catalyzing the beginning of my serious study of our flawed military discipline system; hence, after much research and with the passage of many years of thought arrives this book.

Immediately, the court-martial had engendered in me an urgent need to discover and ferret out precisely how Bitoff was able to pull off his abhorrent action, one completed with neither keen scrutiny nor outside oversight.

In that examination, I hoped to find the dark secret revealing exactly how this one-sided system of military justice was still available for use by indecent men like Bitoff. Further, I then wished to use that new knowledge to prevent a betrayal like Bitoff's from being inflicted upon any other serviceman. Finally, I hoped to find a way to either force Bitoff to admit to his cruel conduct or to force more senior officers to expose Bitoff and hold the bad admiral accountable.

In sum, I hoped to clear my name and continue onward in the career that I had chosen for myself.

Over time, I recognized that God, in His infinite wisdom, had pulled me out of the military; moreover, over time I came to realize that this departure was a blessing. Had I remained on active duty in the navy, I might have been co-opted to become a coercive manager and an apologizing participant in this calculated woke nightmare that today's military has quietly become.

Still, Bitoff's betrayal is never-ending. It stings more painfully every day as I watch military governors just like Bitoff pull the military into the deepening morass in which it finds itself today, a cavernous morass replete with critical race theory (CRT), poor morale, lousy tin-pot leaders, simple rewritings of history, deeply declining physical and mental standards, and abysmal recruitment numbers, all factors which shout out one message to any citizen who is paying attention: "When attacked, we are not ready! No, sir!" The intensity of Bitoff's betrayal simply grows and metastasizes as it is repeated and reinforced every moment by people just like Bitoff. I have concluded that some of these upper-echelon men and women are actually intent upon the destruction of America as it once was—and, what is more, there is no doubt in my mind that some of them no longer wish for the USA to remain the predominant world power.

Military leaders have been very successful in instituting their tyrannical rule of man, and in that wrenching process, they have insidiously replaced the rule of law found in the United States Constitution with the arbitrary rule of unaccountable military men. I may speculate that people like Bitoff find little redemptive or exceptional in or about America, probably because they are interested only in their protected self-promotion and advancement.

Thus, this work which focuses on our intrinsically dishonest system of military justice began as a simple reaction to Bitoff's stark betrayal. In twenty-five years of living with the consequences of John Bitoff's personal assault upon me and thinking about how and why this happened, naturally my tone and tenor have changed measurably over time.

I loved the navy, and I loved going down to the sea in ships. I loved my family and was confident in the opportunity the navy afforded me to provide for them in a meaningful, consistent way. Using good character in the active defense of the United States, I saw the importance of oaths, and in all my daily duties I worked to implement them fully.

Yet, in a flash, John Bitoff took all of that away from me. I felt blindsided! My first reactions were to recover from the loss and then to expose Bitoff to the navy as an indecent actor. I knew what Bitoff had done, and as stated, I wanted to prevent what he had done to me from happening to others. In the early days the struggle was to clear my name, to gather information, and finally, to endeavor to stay on active duty.

Many words and sentiments that I had heard during my early days in training came back to haunt me. Some of them, I had heard repeated regularly throughout my career; some of those ideas had always concerned me, but now I had the chance to study them more closely.

Once again, here is the nexus of the issue: American servicemen are taught to defend our republican form of government with their lives under the Constitution against all enemies foreign and domestic, but servicemen are not themselves protected by that same Constitution. As a good citizen and student of history, this must move one to pose two central questions: Is this truly the best that we can do? And, given the current arbitrary status of military justice, what can be done to improve it?

After I was honorably discharged into a forced early retirement, my next keen focus naturally turned to making a thorough and methodical study of the history of the military's system of discipline. In those early days, now so many years ago, I was attempting to come to my own fresh understanding of exactly how this cruel governmental anomaly within the military had come to exist. The logical imperative at hand was to discover the nature of the core problem, then work with others to reconcile the fracture and perhaps, if fortunate, make the inequitable discrepancy disappear.

Today, I am both saddened and discouraged to report that the history is known and unhelpful, to wit: abandonment of the Constitution was the intended outcome of a very powerful collection of tyrannical actors. I am further saddened and discouraged to observe that, generally, soldiers and sailors of today are usually more interested in themselves than they are interested in their families' diminishing freedom.

Therefore, individuals opting to "engage" in something bigger than themselves in the conduct of civic virtue are being effaced, neglected, or disgraced, thoughtlessly ensnared in our "me first" society.

During my long years of study after discharge, fortuitously it was President Ronald Reagan who pointed me in the right direction when on August 23, 1984, at a Dallas Prayer Breakfast, he said:

> Without God, there is no virtue, because there's no prompting of the conscience. Without God, we're mired in the material, that flat world that tells us only what the senses perceive. Without God, there is a coarsening of the society. And without God, democracy will not and cannot long endure. If we ever forget that we're one nation under God, then we will be a nation gone under.[3]

---

3 Regan, Ronald. (1984). Ecumenical Prayer Breakfast. Original published in Dallas, Texas, August 23, 1984.

However, presaging the second-class status of service members, John Adams much earlier instructs:

There was extant, I observed, one system of Articles of War which had carried two empires to the head of mankind. The Roman and the British: for the Articles of War are not only a literal translation of the Roman. It would be vain for us to seek our own inventions or the records of warlike nations [*sic*] a more complete system of military discipline. I was therefore for reporting the British Articles of War *totidem verbis* [...] *(with just so many words).*[4]

Then, the full force of the idea hit me: the armed forces of the United States stepped away from God at the beginning of conflict under arms from our earliest days, way back in 1775.

The American Articles of War remained in effect through July 2, 1776.

⚓

THE JAGMIRE!

A navy is essentially and necessarily aristocratic. True as may be the political principles for which we are now contending they can never be practically applied or even admitted on board ship, out of port or off soundings. This may seem a hardship, but it is nevertheless the simplest of truths. Whilst the ships sent forth by the Congress may and must fight for the principles of human rights and republican freedom, the ships themselves must be ruled and commanded at sea under a system of absolute despotism.[5]

—John Paul Jones (traditional), September 14, 1775,
Commodore in the Continental Navy

---

4 Valle, James E. (1980). Rocks and Shoals: Navy Discipline in the Art of Fighting Sail. Annapolis, Maryland: Naval Institute Press, pp. 40–41.
5 John Paul Jones letter under date 14 September 1775 to the Naval Committee of Congress.

Thus, we run the military under three forms of government: one, the Roman; two, the British; and three, the American shadow government, none of which operate explicitly under God.

By the way, in this context we must consider the case of *Dynes v. Hoover*. In 1857, a seaman was prosecuted for desertion; furthermore, the case pronounced the importance of US Constitutional oversight. However, over one and a half centuries later, the case has, unfortunately, made little real difference and has only been employed sporadically. In addition to a military criminal act being judiciable under the Uniform Code of Military Justice, such a case may also be raised in civilian federal court under the *Dynes v. Hoover* doctrine, one which establishes federal judicial oversight of court-martials on these two grounds: (1) it did not have jurisdiction over the subject matter or charge, and (2) it failed to observe the rules prescribed by the statue for its exercise.

⚓

During my patient study of this issue, I learned that for someone who, through loss of a capitol ship of the first rank through attack, or someone who, of sound mind and body, espouses a conservative mind and makes thorough attacks on a subordinate, the courts-martial is one of the ways to take out the "miscreant."

Thus, Edward M. Byrne, Navy Captain JAG and author of *Military Law*, informs us:

The courts-martial is an independent court of law [...]. Congress through the Code, has removed as an instrument of the commanding officer desires in any case [...] but once he refers a case to a courts-martial, he cannot influence the proceeding to the accused detriment. The question of guilt or innocence of the accused and the sentence to be awarded in the event he is found guilty is an independent decision of the court.

Article 37 of the Code states:

"No authority convening a general, special, or summary court-martial, nor any officer, may censure, reprimand, or admonish the court or any member of the court or any member, military judge, or counsel thereof, or with respect to any other exercise of its functions in the conduct of the proceedings. No person [...] may attempt to coerce or, by any authorize means, influence the action of a court-martial or any other military tribunal or any member thereof, in reaching the findings or sentence in any case, or the action of any convening authority, approving, or reviewing authority with respect to his judicial acts."

– Edward, Byrne M. (1981). *Military Law*. United States Naval Institute, Annapolis, Md., p. 285.

In a similar vein, William Winthrop, Army JAG, espouses the same notion in *Military Law and Precedents* in his 1896 two-tome doorstopper: "For executing an illegal sentence of a military court. That an officer who executes the sentence of a military tribunal which was without jurisdiction, or whose proceedings or judgement where otherwise illegal so that the sentences is invalidated, is a trespasser, and liable to an action for damages on the part of the person sentences [...]." (Winthrop, William. (1896). *Office of the Advocate General*, Document No. 1001, Washington, DC, pp. 881–882.)

I hope that the kind reader will begin to sense that there are similarities between these two historic courts-martial that did take place, first in 1945 and then in 1989 to 1990. Secretary of the Navy James Vincent Forrestal (1892–1949) and Rear Admiral Bitoff both repeated the same malicious sin: Forrestal called out Rear Admiral Chares Butler McVay III for a hugely unfair punishment and later, Rear Admiral John W. Bitoff stacked the court-martial board in his favor against myself! Engendered in both cases by cruel vitriol, revenge, and scapegoating, neither military trial should ever have taken place.

⚓

Convening Authorities James Forrestal and John Bitoff broke these laws in the two flimsy and fictitious courts-martial of Charles Butler McVay III and Walter Francis Fitzpatrick III. Soon, in coming pages of this book, I shall carry out a step-by-step analysis of both of these courts-martial.

⚓

In June 1970, my plebe year, the United States Naval Academy Honor Code read: "A midshipman will not lie, cheat, or steal."

Later, after a scandal erupted with the discovery of massive cheating during an academic examination, the Honor Code was altered to read: "A midshipman will not lie, cheat, or steal, nor tolerate someone who does."

This book is an exploratory treatise on our unjustifiable, biased system of military justice, one in serious need of immediate examination and careful, judicious reform.

⚓

THE OPERATION OF AMERICA'S MILITARY DISCIPLINE SYSTEM IN MODERN TIMES

The purposes of this book are first to remind us of the many freedoms that we have lost long ago, and then to force or compel their restoration. Thus, this work strives to prove that the military law is an instrument of oppression, that it is intentionally rife with corruption and undue influence, and that it operates as an outlawed government entity, well outside the bounds of any controlling legal authority, populated by the generals and admirals of the upper military establishment. Courts-martial are truly medieval "star chambers" in the relentless dissemination of their clandestine Acts of Supremacy.

⚓

In late 2001, US News & World Report Senior Investigative Reporter Edward T. Pound called Glenda Ewing and me. Mr. Pound had learned of our appearance and testimonies before the 23 March 2001 COX COMMISSION, an indictment of Chinese espionage and lax security standards at US laboratories. At that time, Ed was beginning work on a report on the military discipline system. For the next year, I worked closely with him as a research assistant. Ed's article published as the cover story "Unequal Justice: Why America's Military Courts Are Stacked to Convict" on 16 December 2002. In it, Ed profiled several NCIS abuses.[6]

⚓

However, other books and writings on the tradition, practice, history, and structure of the military establishment's discipline system attempt to defend it. Truth be told, uniformly, those works struggle in tortured, twisting graphs, in wiring diagrams of boxes, arrows, and charts; they use endless convoluted words of explanation that attempt, but ultimately fail, to paint a clear, logical, understandable picture. I offer, dear reader, as an alternative, this work's version that is straightforward, one owning clean and clear imagery.

Please consider George Washington's picture on the dollar bill where, to the left, a pyramid is pictured overseen by a light-emitting "eye," i.e.,

6 Pound, Edward T. (16 December 2002). "Unequal Justice: Why America's Military Courts Are Stacked to Convict." U.S. News and World Report, 133 (23), p.18.

God's watchful eye.[7] Many a military man reading these words will nod and chuckle their agreement to this diminutive allegory, stating that, when applied to any military organization, only the commanders are gods.

Please think of our military government as a gargantuan pyramid scheme. At the top resides the president of the United States, our commander-in-chief, who Congress singularly invests with enough statutory power to punish every person populating the military establishment below.

The president delegates his authority to subordinate commanders just underneath in the pyramid, all perched at or near the top, empowered by the president to discipline those below, but not above.

In ancient times, such pyramid-style governments were the province of kings, queens, pharaohs, emperors, and empresses in the administration and exercise of their military war-making powers. They wore helmets instead of crowns in their roles as marshals,[8] generals, or admirals taking up arms, wielding shields and swords.

⚓

Today, unfortunately, military law works in the same way it once did all the way back in its "arcane"[9] roots in antiquity.

Those who argue that military law is in any way constrained by the United States Constitution are exposed and condemned. Shortly, the courts-martial convened by Navy Secretary James Forrestal and Navy Rear Admiral John Bitoff will be profiled in detail to drive these points home.

---

7 One of America's mottos, *"Annuit cœptis"*: He (God) has favored our undertaking bends around and above the sparkling eye. A second motto appears wrapped at the bottom of the pyramid: *"Novus ordo seclorum"* translates from Latin, "a new order for the ages." This new order began wit.
8 A military rank.
9 Appendix One: As John Bitoff relates in his 30 April 1999 letter to Congressman Norm Dicks.

This work features the following four focal points:

Point One: Of primary interest is the operation of a non-Constitutional military government within the United States. As stated, this second government was born of a Roman/British scheme of discipline and neatly fits George Orwell's *Animal Farm* model: quiet dominion over the helpless and hapless.

Point Two: America's military governance is maintained through the complex practice of attainder, pains, and penalties, one replacing America's constitutional court trials. The Ancient Roman, then British system of attainder is structured to resemble the US federal court trial organization and procedure through the hands of Benjamin Franklin. In truth, that system of military governance is the operation of a modern-day monarchy.

Point Three: Essentially, two kinds of officers exist in today's US military establishment, and in Anton Myrer's 1968 novel *Once an Eagle*, they are symbolized by Sam Damon and Courtney Massengale acting as protagonist and antagonist, respectively. Sam Damon stands for the kind of officer who leads from the front and who patently rejects the abuse of subordinates by any means, most especially through the practice of attainder—that is, courts-martial. Opposite him, we see Damon as the military politician. "Court" Massengale is someone who is willing to sacrifice anyone who might obstruct his path to glory and promotion. For the Court Massengales of America's armed forces, the use of attainder by courts-martial is a particularly functional tool, and one that is employed frequently.

Lastly, Point Four: America's military government must be abolished as a matter of constitutional mandate, and, in turn, rightly reconciled with the Constitution so that men and women in charge—such as Secretary James Forrestal and Admiral John Bitoff, indecent commanders—are exposed to the consequences of their criminal acts as any other American would be.

Today, in the exercise of military discipline and punishment, or the courts-martial system, military sovereigns (government officials, generals and admirals, and flag officers) like former Navy Secretary James Forrestal and Rear Admiral John Bitoff are invested with unrestrained partisan power. Even in this modern day, they are endowed with an ancient, arbitrary, excessive, and summary war authority over the good names, characters, freedoms and liberties, and the very lives of those under their command and their families. This dire and unfair situation cannot remain.

The antics of Ancient Rome (including Caesar Augustus, among many others), Britain, and then America eventually merged with the administration of the Army of Britain. Finally, the slapping together of Benjamin Franklin's hands marked the creation of the shameful military justice system that the US Army, Navy, Air Force, and Space Force all employ today.

Further help in the construction of this biased system was afforded by the direction of Army Colonel William Winthrop, Marine Corps Colonel Brigadier General James Snedeker, and Navy Captain Edward M. Byrne, all of whom are contributors to the fuller structure in which we find ourselves today. Therefore, it is natural and fitting that I refer to these men as "command racketeers."

⚓

Collected here, in one place and in devastating, detailed, and crucial context, is a book covering our current flawed military justice system. It is made available to the larger public for the first time, and it is intended to ignite a modern-day polemical debate centering on the scope and operation of America's military discipline system: the court-martial system.

Thus, the two most sacred assignments that I shall carry out in my life are:

Assignment One: To offer a key to open the door to civilian judicial oversight by the federal (nonmilitary) criminal courts.

Assignment Two: To make the navy and marine corps, the Naval Criminal Investigative Service, the FBI, and Congress "look really bad!" That is, to bust the phony image and make clear and lucid the hazy mirage.

Thus, a fresh examination, inspection, and debate focused on bringing our military government into agreement with the US Constitution is long awaited, and long overdue.

It is an ironic military maxim that US forces must take an oath to defend the Constitution with their lives, yet they themselves are not protected by that very same document. Very few citizens in our nation can explain why this incongruity continues to exist.

Again, we shall briefly detail it here: It is the triumvirate of three sovereigns—the Ancient Roman, the British, and then the shape-shifting American governance. It is "lawfare" writ large, with all three crafted in the absence, rather than in the presence, of God. As Pete Hegseth of Fox News tells us: "Without God we are prone to power consolidation. He is the architect of our good fortune" (September 16, 2022). How could a system of military discipline constituted without God ever exist without rancor or virulence, without corruption and overarching power grabs? Along the way, we must ask ourselves the salient question: Was this unhealthy triumvirate of coercive forces planned out, or did it just spontaneously happen?

This book shall offer to the reader a suggestion that empowers them to become the catalyzing function behind fresh, much-needed oversight, review, and repair of the current system of military justice and, furthermore, full reconciliation of that system with the Constitution.

To this end, in coming chapters we shall focus precisely on two fundamental military trials: the James Forrestal court-martial (November-December 1945) and the John Bitoff court-martial (1989–1990). These two painful events supply extraordinary revelations and clearly expose our system of military discipline as a corrupt organization run by command racketeers.

The history regarding the Forrestal and Bitoff courts-martial demonstrates the naked arrogance of both civilians and uniformed criminals in command alike.

These two courts-martial read like a subdued step-by-step mystery novel. Powerful passions will be evoked along the full reach of emotions. The two sad stories are dynamic tales grippingly told in a dramatic setting.

Using the research garnered by my own independent investigation, augmented by voluminous files and papers amassed over the course of nearly a quarter century and then relentlessly scoured for key details, this work details the wanton avarice of rank and the unmitigated abuse of power that continues to this day within all branches of the military.

I stress again the outright rejection of due process as a fundamental right denied to members of our Armed Forces. The gentle reader shall soon discover that the military's judge advocate general (JAG) priesthood has thrown acid on our sacred Constitution. I shall explain how our Congress and top Defense Department personnel routinely turn a blind eye in their official and unofficial approvals, thus giving their tacit sanctions to all types of unfairness, one-sidedness, and iniquity.

The military's unique governance of discipline and punishment is solely the function of command, of juniors reporting to seniors, of juniors obedient to the orders from above, and of juniors particularly vulnerable to the outrageous conduct of their superiors. That governance lacks any type of proper checks and balances, any sort of nominal brake that would stop the military from exerting too much power.

Therefore, there is nothing particularly "judicial" about America's military government under the Uniform Code of Military Justice (UCMJ). Within—painstakingly detailed, using hard-hitting facts one after another after another—I hold that the UCMJ is merely window dressing and flimsy artifice. Indeed, it is commonly used to hide in total darkness the very serious real-world abuses visited upon US servicemen, often reflecting the capricious whims of the one in command. Thus, if ever properly tested against our US Constitution, command influence would be quickly and starkly revealed as a mortal enemy to our American and Republican form of government.

Clearly, the abuse of our military members that has persisted through our history has evoked and continues to evoke powerful passions. In this atmosphere, where generals and admirals in the right position of power within the pyramid can do virtually anything, how many thousands of service members have had their rights truncated? What are the long-term ramifications of our flawed military justice system upon the military itself, and its very idea of leadership?

Still, to today, unless corrected, the aristocracy that runs our military remains willing to let innocent service members and their families suffer the burden of wrongful federal convictions, rather than condemn those officers who abuse their power. That aristocracy uses the military punishment regime as a political tool, as a cloaking system, and often as a weapon of cruel vengeance.

Today, military authorities stand not only above the law, but also completely out of reach. In their rarefied air, silence is acceptance; that arrogance is abuse. Nonetheless, many of the military authorities named here within this work remain criminally accountable.

Walter Francis Fitzpatrick III was an officer who had the courage to speak out for his men and defend them against the capricious vagaries of a wrathful senior. For his trouble, those military authorities took the law into their own hands and used the military discipline system to destroy an innocent man who stood by the courage of his convictions. And for that courageous action, he was convicted!

I hope that the reader will come to realize why the military government must finally submit to a full constitutional reconciliation. It is my firm hope that in time, congressional leaders will come to demand that desperately needed reconciliation!

Therefore, unfair and capricious command influence must once and forever be aggressively condemned and eradicated.

Several individuals who hold appointed and elected federal government positions perjured their oaths and refused to pledge their allegiance to America. They are bent on destroying our republican and constitutional form of government.

These same government officials of all types and kinds are criminals at large. Many have committed serious crimes, and yet they receive no handcuffs, and no prison time.

Accordingly, now is the proper time to demonstrate in detail to the reader exactly how an arbitrary system of military discipline operates.

John Bitoff and his criminal assistants never expected for his illicit adventures to be publicly exposed. Thus, much of the embarrassing and illicit communication that follows was never intended for any sort of public review.

This analysis of the contorted military justice system cannot be found elsewhere. The document record contained herein has now, for the first time, been published in book form.

One must ask the question: What is to be done when this extraordinary power is vested in a man like John Bitoff, who then inflicted untoward power so outrageously on innocent subordinates?

But there remains a wider audience to whom I have more to report. Charles Dicken asks the pertinent question, as a civilian, in his novella *Hard Times: For These Times* (Bradbury & Evans, 1854) here paraphrased: "What happened to the laws that are supposed to protect us?"

In the American military culture, the question is more complicated, and once again the irony is posed: How is it that US servicemen are sworn to protect and defend the Constitution with their lives, and yet they are not protected by that very same Constitution? Once again, a good citizen must ask the quite obvious question: Is our present situation of military justice the best we as a people and nation can do?

I hold that I do answer both questions satisfactorily with this work.

After the long and difficult trial that I have lived through, naturally and deeply I began to learn about and study William Winthrop, whose writings, as we have seen, underpin the entire system. I needed to understand exactly how an innocent man could be brought through this wrenching court-martial process, and the precise, and ugly, manner in which both Forrestal and Bitoff pulled it off.

It is my firm intent that for most readers, this study, like none other, shall engender and support only clear contempt for William Winthrop's weak, flimsy, and biased apologetics for the American military discipline system.

Based upon my findings, I do demand, as countless others before me have also done, a full review of uncounted courts-martial convictions and, after that diligent action, the systematic overturning of many.

This work slowly and methodically uncovers the full scope and operation of America's military discipline system, a unique and tortured system of government.

A document record of the inner workings of a court-martial is nowhere else published. Here is the obvious nexus, dear reader: In the absence of independent oversight, the workings of military discipline, without the Constitution as a counterforce, shall naturally be fraught with all manner of abuse, vengeance, and corruption. If irked or provoked, commanders are free to execute their acts of supremacy upon subordinates without fear of independent oversight from the likes of a grand jury or federal judges. Normal judicial review, common elsewhere, simply does not exist here. In such a godless environment, evidence is suppressed, documents are doctored or destroyed, charges are inflated and exaggerated, false statements are manufactured and entered into the records, witnesses are invented—and for all of this, no consequence befalls the abusive commander.

Diverse results and ramifications from centuries of this unfair military justice have followed and are now visited upon the entire nation. In time, evil actions produce dire consequences. Since an unfair justice system has been in place for centuries, an improper model for leadership has emerged and become common. In this atmosphere of automatic cronyism put in place by the military, it follows that career advancement takes place not due to performance, but, primarily, by those skillful and adept at pleasing their bosses. That is not a good thing! In the military, it used to be that simple performance merited advancement; in other words, a strict meritocracy once was in place. Yet today, sycophants rule, and in this foul, self-aggrandizing atmosphere, rapidly they climb up the ladder. As a result, a spreading wokeness like a cancer reigns, not just in the military but in business generally. Therefore, military leaders of today promote conformity over completion, rear-end kissing over achievement, and doormat yes-men over true, decisive leaders.

On the other hand, in his book *The Generals: American Military Command from World War II to Today* (published by Penguin Books in 2013), Thomas E. Ricks speaks about how in World War II, a military officer had only ninety days to achieve a key strategic objective, or else he would be summarily replaced. How far we have drifted away from that rigid, correct standard! Today, sadly, the motto is "To get along, go along," or "One conforms to garner and coalesce acceptance and security." Today, instead of focusing on genuine standards of leadership, we have allowed the fawners and flatterers, the doormats and flunkies, the leeches and suck-ups to take over. These are just a few of the long-term ramifications of the unfair system of military discipline that, unfortunately, has been in place since our nation's founding.

As this prologue happily draws to a close, the writer must mention that King Henry VIII inspired the book's title. In 1534, Parliament passed the Act of Supremacy that defined the right of Henry VIII to be the supreme head on earth of the Church of England, thereby severing all ecclesiastical links with Rome. It is not a good idea to turn away from God!

Rather than a theoretical polemic, this book primarily relates a real-world, jarring example of how the unconstitutional court-martial machine works daily to inflict injustice and to grind up people in service to military discipline.

Further, this endeavor works to carry out repairs to my father's service record as well as my own.

Clearly, one step in the path forward is to expose those involved in the James Forrestal court-martial and the Bitoff-Zeller-Anderson court-martial and to stand them all before a civilian federal judicial remedy (e.g., *Dynes v. Hoover*, 1857) as well as the federal criminal prosecutions that must transpire after all these long years as criminal accountability. And that is because all these men, living or dead, must finally face their criminal accountability.

The reader is invited to follow along on this firsthand exposition and to read the sad records for yourself.

Thus, please read about these two courts-martial—the James Forrestal as convening authority, and the Bitoff-Zeller-Anderson as convening authority—and then, in both cases, ask yourself whether justice was served and whether a correct, faithful outcome was attained.

So, welcome aboard, as we are about to get underway. Bravely, let's sail into this nasty fray that must be joined!

Much of our journey will involve sailing through foul weather and heavy, stormy seas. Stand on board the ship with your feet spread wide apart so that you do not get tossed into the sea. Indeed, this voyage may make you sick! So, have your seasickness remedies handy as we set the special sea and anchor detail and proceed to open ocean to meet first-hand the terrorist group "November 17" that first formed in 1975.

Underway, shift colors.

Beware.

⚓

Dr. Victor Frankl instructs, "[...] there are two races of men in this world, but only these two—the 'race' of the decent man and the 'race' of the indecent man. Both are found everywhere [...]."[10]

Of course, the challenge for every reader is to sort out the descent from the incident.

For consider deeply, dear reader, all the other soldiers and sailors who have had to suffer the grinding process of a court-martial—Army Lieutenant Clint Lorance, Army First Lieutenant Mark Basham, Army Lieutenant Colonel Terrance L. Lakin, and Marine Lieutenant Colonel Stuart Scheller, among others—and the suffering that could have been avoided if an effective oversight system were in place.

Finally, these many words of prologue have been written by Survivor #5 of the Desert Duck helicopter crash, which took place at 17:58 hours (local) on 30 July 1987[11] in the central Persian Gulf.

10 Frankel, Victor E. (1959). Man's Search For Meaning. Boston, Massachutes, Beacon Press, pp. 81, 179.
11 Survivor Says Helicopter Came to Low. Unsigned, (1987). AP news. Retrieved from https;//apnews.com/article/51d3131984d60af84a23776b7d5cc876. Note: I crashed into the sea the same day as the USS Indianapolis (CA-35), twenty-two years later.

1.
BILL AND MIKE

"The highlight of my childhood was making my brother laugh so hard that food came out his nose."

—Garrison Keillor (1942–Present)

"When brothers agree, no fortress is so strong as their common life."

—Antisthenes (445 BC–365 BC)

⚓

"ANCHORED! SHIFT COLORS!"

"Secure from the special sea and anchor detail."

"Set the anchor watch."

The *USS MARS* rode peacefully at the hook after executing a textbook-precision anchorage in San Diego Harbor, a short boat ride from the Fleet Training Group Headquarters ashore.

The navigator and his quartermasters took a round of bearings used to triangulate and fix the ship's position. The first lieutenant standing with his boatswain mates on the fo'c'scle reported how much chain was paid out. A swing circle was struck on the chart. The navigator and his quartermasters shot their first round of bearings, then drew lines from landmarks ashore to the ship to triangulate and fix our position.

The bridge anchor watch repeated this standard operating procedure every fifteen minutes, as the anchor was never completely trusted to hold us fast.

Other bridge watch standers took off their sound-powered phones and stowed them away, ready for immediate use. Moving off the bridge in small groups, they struck below to attend to their primary and varied shipboard duties.

For the engineers down below, little changed in their watch-standing duties. The main propulsion plant operated at anchor as it operated as if we were sailing upon the open ocean. In the event the quartermasters discovered the ship was not on or had drifted outside the swing circle, meaning we were dragging anchor, the sea and anchor detail would be reset in an instant. Orders would be issued to the main space, using the main engine to hold our position and power the auxiliary gear necessary to recover and re-drop the anchor.

Normally, our operational commander in the Third Fleet was a one-star admiral headquartered in our homeport aboard the former Naval Supply Center in Oakland, but there comes a time in every combat vessel's life when she must be tested and evaluated in her ability to fight.

Evaluating and testing a ship in its ability to effectively engage and defeat an enemy force is, at its core, a very specialized kind of performance audit. It is the kind of examination best carried out by people other than those whose fitness reports and future careers are not affected by the results of the evaluation.

As soon as the anchor struck bottom and took hold, the command structure for the ship changed. In other words, Mike Nordeen had a new boss. In that instant, the *USS MARS* was no longer operational in the combat logistics force, no longer part of the Third Fleet.

Careers are enhanced or harmed based on the performance of a ship in refresher training. Lessons learned in the attack and near sinking of the *USS STARK* (FFG-31) in May 1987 had dramatically affected what were already rigorous and difficult performance requirements.

It ought to be stated that ships routinely fail their combat performance audit.

Naturally anxious, our skipper, Navy Captain Mike Nordeen, paced about on the bridge as it emptied out, walking about for a bit before sitting down in the portside bridge chair.

On that day, I was Mike Nordeen's executive officer. Worried more about Captain Nordeen's state of mind than how we would do during our training inspection and exam, I hovered on the bridge with the skipper.

Yet, pretending to be relaxed, the skipper had his legs crossed at the ankle. They were kicked up onto a structural strength member angle-iron beam, part of the forward bulwark bulkhead. He was focused, gazing out in a ten-thousand-yard stare—not really looking at anything, just thinking.

Our pre-deployment scheduled had been compressed and unnaturally frenetic. We had been delayed in getting back to sea due to an unexpected and lengthy main shaft repair effort. Normally, we would have participated in operational fleet exercises; those exercises were very important to captains and crews who used these at-sea fleet maneuvers to practice diligently, always with an eye toward the final battle problem exam we would we need to pass before overseas deployment: Refresher Training.

We would also have had opportunity to operate, check, and finetune our equipment and train for an engineering propulsion plant exam, which was but another hurdle in making ready a warship intending to sail into harm's way.

I approached to the captain's left, knowing what was on his mind, and told him we were ready in all respects. Further, I added that the crew was as skilled as they were motivated, and finally, that we were going to excel.

The captain nodded in agreement, as he had so many times before. After all, we had had this same conversation on several occasions.

For no other reason than to gently force Mike Nordeen to think about something else, I asked him how it was he had come to join the navy.

He turned his head and smiled, taken by surprise. Clearly he had not been expecting the change of subject.

It was then that Mike Nordeen first told me he had an older brother, William. Next, the skipper explained why he routinely and affectionately described our vessel to others as an "attack food ship."

For one brief moment, Mike Nordeen was back in Centuria, Wisconsin. Growing up in that isolated village in the northwestern corner of the state, Mike and Bill had worked together in a small neighborhood grocery store that their father owned. Captain Nordeen enjoyed reminiscing about how he and brother Bill would stock the store and put up for display fresh foods and vegetables.

The skipper remarked how he was right back at home and in his childhood element now that he was in command of one of the navy's "attack food ships," dubbed an AFS.

Older brother Bill had later left Centuria to join the Navy.

Bill had successfully trained as a pilot and earned the coveted gold wings of a naval aviator, learning to fly helicopters. In turn, Mike soon followed and trained in jets. As the Nordeen brothers progressed in their careers, both men were groomed to assume more senior non-flying staff and command postings and, working diligently, both were subsequently promoted to the rank of navy captain (O-6, the sixth highest maritime officer rank) by 1988.

Bill Nordeen, in June 1988, was the military attaché to the US embassy in Athens, Greece. Captain Mike explained that his brother Bill, Bill's wife Patricia, and their twelve-year-old daughter Annabel were just three weeks away from returning to the United States to enjoy a planned retirement culminating Bill's thirty-year navy career.

The USS MARS was the first of a new class of ships assigned the designation "AFS"—naval speak for "combat stores ship."

The AFS replaced three earlier types of supply ships, combining three separate afloat combat stores, missions, and capabilities into a single vessel: the AF (store ship), AKS (stores issue ship), and AVS (aviation supply ship). The AFS was capable of carrying and delivering supplies to the fleet steaming in a combat zone, everything from micro-miniature repair parts to ten-thousand-pound jet engines.

The *USS MARS* was the first Pacific Fleet ship equipped with the UNIVAC 1104 computer system used in the tracking and management of over one million line items of supplies stocked onboard. The *USS MARS*, as three ships rolled into one, was, in fact, a Naval Supply Center afloat.

The *MARS* delivered a great deal more than just food.

Mike Nordeen and Bill Nordeen were close. It showed.

Mike Nordeen followed in the wake of the guide, while Bill Nordeen flew helicopters.

Mike Nordeen was a fixed-wing attack jet fighter pilot.

It was a glorious day as we watched the sun go down on the western horizon, the type when the weather delights the Chamber of Commerce folks who promote San Diego. That day on board the *USS Mars*, the skipper told me how Bill was extremely tense and on high alert because of the bloody history of the uniquely Greek terrorist group calling itself 17 November (to be tabbed N-17, or 17N).

Bill Nordeen was on pins and needles, acutely aware that N-17 had proudly claimed credit for the assassination of another navy captain five years before in February 1983. Captain George Tsantes Jr. had held the very high-profile position of deputy chief of the Joint US Military Advisory Group in Athens, Greece, when N-17 riddled his car with a hail of gunfire, thereby killing both Captain Tsantes and his driver.

We talked a little more.

The skipper chuckled when he recalled how both brothers had married women named Patricia.

The sun dropped below the horizon. It was getting chilly on the bridge.

I assured Mike Nordeen once more that we were ready to go. The crew was going to knock this refresher training drill in the head. I told the captain I was going to make a quick tour of the ship before I returned to my stateroom, did some paperwork, and then hit my pit.

With a sharp salute, I requested permission to lay below.

"Permission granted, XO."

The bridge was empty save for the quartermaster on the anchor watch and the skipper.

I turned, walked to the navigation table, checked the chart to find the ship resting comfortably within its swing circle, and finally departed the bridge.

As I toured the ship, I more carefully considered, without having asked myself the question, that the cause of Captain Mike Nordeen's discomfiture had more to do with praying for his brother's and his family's safe return than any nervousness about the coming REFTRA performance. He knew that his outstanding crew would do well on that training inspection and exam.

Still, we were all tired. There was a great tension in the air when I hit my bunk.

We were ready for refresher training.

But by no means were we ready for what tomorrow was about to offer.

2.
THE ELUSIVE ASSASSINS OF ATHENS!

⚓

BILL NORDEEN, US EMBASSY DEFENSE attaché, arose on the morning of Tuesday, 28 June 1988, dressed, and had breakfast with his daughter, Annabel.

He left the house located in the suburb of Kefalari and walked to his armored-plated Ford Grenada sedan, courteously acknowledging the uniformed Greek police officer standing watch and guarding the Nordeen residence. Captain Nordeen started the Ford and started out to work at 8:00 a.m.

The only ingress-egress road was lined with a column of arbors on both sides. It was a narrow tree-shaded path. Three hundred feet down this road, away from the Nordeen residence, a Toyota was parked.

Only minutes had passed since breakfast. Bill drove down the narrow road on the single street leading out of his community.

Lurking somewhere in the area was a member of N17 holding a detonator, thumb on the trigger.

As Nordeen passed the Toyota, it erupted in a fireball, shooting flames fifteen feet into the air.

Force from the explosion threw Captain Nordeen forty-five feet from his heavily armored Ford vehicle into the yard of a deserted villa.

Windows were blown out for blocks away.

Annabel came rushing out of her house and ran down the street to find her dismembered father.

N-17 operatives had packed the trunk of the Toyota with fifty pounds of dynamite and plastic explosives.

Athens, Greece, is ten hours ahead of the local time along California's west coast.

It was 10:00 p.m. Pacific Daylight Time on the evening of Monday, 27 June 1988, when Bill Nordeen lost his life. The *USS MARS* was at sea, sailing south off the west coast toward her San Diego anchorage.

While my skipper, Mike Nordeen, and I reminisced on the bridge of the *MARS* about his childhood recollections of the two brothers growing up, Bill Nordeen had already been with our Father God for around eighteen hours.

News of the terrorist assassination of Bill Nordeen did not make it to the ship for approximately twenty-eight hours, coming in on a shipboard teletype around 2:00 a.m. Wednesday morning, 29 June.

Mike Nordeen, commanding officer of the *USS MARS*, was sleeping aboard ship at anchor in harbor in San Diego, California, when a highly sensitive message, transmitted to military commands worldwide, clicked off the radio room teletype—one announcing the terrorist attack in Athens and the murder of Mike's brother.

The duty radiomen on watch on the *USS MARS* received the highly classified and sensitive message in the very early hours. Reading it, the watch stander, running his eyes over the name Captain William Nordeen, reacted quickly and with alarm. The radioman, obviously nervous that his commanding officer Captain Mike Nordeen and the assassinated Captain William Nordeen might be related, sent a messenger to awaken the command duty officer (CDO).

CDO Charlie Andrews did not know, either.

It was about 03:00 hours local time. Sensitive and highly classified message in hand, the CDO and duty radioman came to my door and began knocking and calling to me with the sense of urgency that only signals bad news. Mike Nordeen's stateroom was just forward and to starboard of mine. The two watch standers were, to a degree, hushed in their attempts to wake me, not yelling, screaming, or banging on my door so as not to wake Captain Mike Nordeen, who was sleeping close enough nearby to hear.

I jumped up immediately, thinking the ship was on fire, flooding, or both—or that someone had died.

I read the message and understood its meaning in a flash. It had been only a few hours prior that I had first learned Mike Nordeen had a brother in the navy stationed in Greece.

And now I learned of Bill Nordeen's demise.

I directed the duty radioman to return to Radio Central and make contact via secure voice radio with watch standers ashore at the Naval Surface Force at Pacific Fleet headquarters in Coronado, California (very near the USS MARS anchorage). I would be up in a moment to speak, intending to confirm the information contained in the message, and upon confirmation we would notify high command regarding the relationship between our commanding officer and the murdered Bill Nordeen.

I told Lieutenant Andrews to go below and get Lieutenant Bradford (Brad) Ableson, our command chaplain, and bring Brad up to my stateroom. Both men were to wait for me there. I got dressed and proceeded to Radio Central.

It took a full forty minutes to exchange information with folks ashore, and sadly, the result was that Bill Nordeen's murder was confirmed. The fact that Bill Nordeen's surviving brother was the commanding officer of the USS MARS, presently assigned to the Fleet Training Group at anchor only miles away, was similarly conveyed. I explained that our chaplain and I were about to notify Mike Nordeen and asked the watch standers ashore to contact the personnel on duty—the Oakland Headquarters Combat Logistics Group 1, our normal operational commander when not engaged in REFTRA. I also asked that the commander of the Fleet Training Group be notified.

I planned to further inform the Fleet Training Group (FTG) senior officers by telling the FTG instructors when they arrived on the *USS MARS* later in the morning.

Chaplain Ableson was in my stateroom when I got back. I told Brad what I knew, and we briefly discussed how to break the news to our skipper. We left my quarters and paced the twenty feet or so to Mike Nordeen's stateroom door.

I knocked firmly. Waited. Then I knocked again.

Pilots—most especially military pilots—are used to death. They know that they are shaking hands with the grim reaper every time they strap on an aircraft and launch. Unfortunately, untimely death in the naval aviation community is both commonplace and expected.

But death always comes knocking on your door when you least expect it.

Mike Nordeen opened the door.

It was about 04:15 hours.

As soon as Captain Nordeen saw me and Brad, he knew we were there to disclose very personal and undoubtedly bad news.

"Who is it?" the skipper said, stoically and laconically. "Pat? My kids?"

Mike Nordeen was half dressed. He was buttoning his khaki shirt, looking right at Brad and me as he asked his questions. Then he turned and walked to his desk, tucking his shirt into his pants while Brad and I followed.

"It's your brother, sir," I said.

I followed the captain to his desk. The captain sat down, and then Brad sat down. I handed the message to Captain Nordeen and took a seat next to Brad, with the chaplain sitting closest to our commanding officer.

I explained to the skipper that I had alerted higher command and they had confirmed the report Mike Nordeen was holding. There were no more details available at the time beyond what was contained in the naval message. Our captain was visibly relieved no one in his immediate family was involved, but still gravely distressed and quite stoic.

Then Captain Nordeen said, changing the subject to matters of immediate shipboard operations, "I'm staying with the ship."

Again, today was the first day of our mandatory pre-deployment operational battle problem testing exercises to be administered by the Pacific Fleet's operational testing organization. Refresher training remains to this day a very serious and important part of preparing ships to go in harm's way.

I gently responded that watch standers on the beach had indicated that Mike Nordeen should be prepared to leave the ship, refresher training notwithstanding. I agreed with that assessment; the personal dynamic of family was just too powerful. The plain fact that Mike and Bill Nordeen were brothers, both navy captains, both serving simultaneously on active duty, was reason enough to temporarily relieve Mike Nordeen of his duties. This way, he could better attend to the needs of his extended family and also act as his brother's escort on the return trip home.

Additionally, it was foreseen that Bill Nordeen's assassination at the hands of a foreign terrorist group would have national and international implications. After all, Bill Nordeen was the second navy captain "November 17" had murdered inside of three years.

Still, Mike Nordeen was adamant that he was going to stay with the USS MARS through refresher training.

Moreover, there was to be no liberty for the crew after we anchored Tuesday afternoon in San Diego.

There was no way any crew member could casually come to know about breaking news from Greece, cloistered as we all were within the lifelines of the MARS. No ship-to-shore personal computer communications existed in 1988. There were no emails, no smartphones, no Facebook, no Twitter—none of it.

What is more, we had no commercial television connections, either.

Commercial radios were used onboard and were fully operational; however, news of the November 17 attack had not yet hit the airwaves giving out names.

On Wednesday morning, only a few *MARS* personnel knew about the assassination of Captain Bill Nordeen.

Mike Nordeen knew. The chaplain and I knew. The command duty officer knew. The Radio Central watch standers knew.

But no one else aboard the *MARS* knew.

When the Fleet Training Group trainers came out the ship the next morning, they carried with them a stack of the area's morning papers as a courtesy gesture for the crew.

San Diego is a devoted navy town. Thus, it was unsurprising that the news of Bill Nordeen's murder had made its way to the front pages by Wednesday morning, June 29.

Training personnel who had come from the beach did not realize the connection between Commanding Officer, *USS MARS*, Mike Nordeen and the announced deceased navy captain in Athens, Greece, William Nordeen.

The shockwave hitting our crew and the visiting trainers lifted the *MARS* out of the water in a flash.

It was our early 9/11 wake-up call, as we experienced it, but these raw emotions and reactions took place way back in 1988.

For the remainder of the day, no one knew what else to do but their duty. We "turned to" focusing solely on the day's schedule. But that schedule, soon enough, was to be modified.

3.
HOMECOMING

Marine Colonel: "Are you related to the deceased?

Army Sergeant: "Yes, sir. He's my *brother*."[12]

⚓

WE SET THE SPECIAL SEA and anchor detail, weighed anchor, and proceeded to the open ocean.

Our trainers had begun to run us through some rudimentary seagoing drills and exercises when word came to the ship that Mike Nordeen had been ordered to Athens, Greece, to attend to his brother's remains and escort Bill Nordeen and his family back home. Captain Mike Nordeen was to disembark the *USS MARS* and proceed directly to Greece.

Rear Admiral Robert Tony was commander of the supply ship group at the time. Captain Michael B. Edwards was Tony's second in command, or chief of staff. Rear Admiral Tony ordered Mike Edwards to relieve Mike Nordeen as commanding officer of the *USS MARS*. Tony dispatched Edwards to travel to San Diego from Oakland and take command of the ship.

At the same time, our crewmembers set about, without a word to anyone, preparing for a memorial ceremony to pay tribute to a fallen brother before our captain departed.

⚓

12 *Taking Chance*. Kevin Bacon.

CEREMONY AT SEA

The men fashioned a wreath.

I asked Master Chief Fa'aita to engage Captain Nordeen in his stateroom to capture and hold his attention.

While he was occupied, our ship's company quietly manned the rail in working uniforms.

A call to the captain's stateroom alerted the master chief that we were ready. Captain Nordeen's presence was requested on the bridge.

The ship had come to full stop and was dead in the water.

It was warm. The sea was glass. And all was quiet and still.

The bridge was silent when the two men arrived on deck. Master Chief Fa'aita escorted Captain Nordeen to the port bridge wing whereupon the skipper stood alone, looking down upon his crew, facing outboard, standing at rigid attention in working uniforms.

Chaplain Ableson read briefly read from scripture. The first verse of "The Navy Hymn" was played on the ship's public address system.

Then, the command was made: "Ship's company, hand salute."

Taps was played as the sailors holding the wreath and the honor guard stepped slowly to the port lifeline; then, it abruptly stopped, and the honor guard dropped the wreath into the water. The group stepped back, came to attention, and saluted.

When the last note of taps sounded came the command "Two," ordering the ship's company to drop their salute.

"Ship's company... dismissed."

On the bridge: "Engine ahead full, right standard rudder."

As the men dispersed, the ship turned toward San Diego.

⚓

CLG-1 Chief of Staff Mike Edwards's helicopter soon approached. We came to flight course and speed, landed the aircraft, received Captain

Edwards aboard, launched the helicopter, and then resumed a course back into San Diego with a full bell sounding, as an important officer would soon depart for Flight Quarters.

Mike Nordeen and Mike Edwards conducted their turnover in the commanding officer's stateroom office. The mantle of command was passed unceremoniously in private, and just like that, Rear Admiral Toney and the Combat Logistics Group ONE staff was short one chief of staff, and the crew of the USS MARS gained a new commanding officer.

A fleet training group (FTG) boat approached and received alongside as soon as we dropped the hook. Captain Nordeen rode back in with our FTG trainer with little more than the shirt on his back.

I was not aware of the arrangements made for Mike Nordeen's travel short of being told he was being jetted directly, travelling nearly nonstop to Athens. Many observed Mike Nordeen as he scampered down the access ladder in his working khaki uniform carrying a single bag. Word got around that he was traveling directly overseas.

Most crewmen on board the USS MARS became unsettled watching Captain Nordeen take off. To wit, something was not right, or something was missing. A nagging and persistent notion filled the air that more action and thought of some sort was needed to support our beloved Captain Nordeen; however, we were not exactly sure what that something was. We were all stunned, almost bewildered, as we struggled to realize just what had taken place. We were dealing with a particular crisis that had never occurred to a ship's crew before—nor, incidentally, has it happened again since.

Unfortunately, there was too little time left in the day after our return to port for deep contemplation. I was personally concerned that the men were unfocused regarding our operational duties by a building and blinding anger, coupled with their sudden shock. The discontented atmosphere on board our ship was heavy and pervasive.

Then, First Class Petty Officer Mark Collins approached Master

Chief Fa'aita with this question: Was Captain Nordeen not going to require more formal uniforms in the days ahead? Beyond that basic concern, more questions quickly emerged about what else Mike Nordeen and the Nordeen family might require by way of support, and this next logical question soon followed: Who was going to provide this support?

William Nordeen had been a member of an embassy staff under the State Department. He was not connected directly to a military outfit.

Arising from these thoughts, there developed a shared and growing insistence that our ship's crew should stand with Mike Nordeen and his family, assisting them through this dark time of terrible tragedy. We sought to represent our ship itself and, with our presence, we would represent the larger navy family as well.

Petty Officer Collins, after further discussion with the first class petty officer's mess, approached the command master chief with the idea of creating a support team for Captain Nordeen. To them, the image of their commanding officer scampering down the ladder all by himself to escort his slain brother home was just too much to bear. This sentiment then grew after further discussions with senior petty officers, who then in turn prompted Petty Officer Collins to suggest to the command master chief the sending of a support team.

Details of Brother Nordeen's assassination spread through the crew quickly and weighed heavily upon us all. Master Chief Fa'aita saw merit in the idea and thought a *USS MARS* contingent traveling east was the right thing to do—but he understood right away that such an idea had to be run up the chain of command.

That Thursday morning, 30 June 1988, the *USS MARS* was once again promptly underway at 08:00 hours.

Seizing upon Petty Officer Collins's idea, Master Chief Fa'aita hastily and quietly called an unscheduled meeting of the *MARS* MWR Committee. MWR stands for morale, welfare, and recreation. The Navy MWR Division administers a varied program of recreational, social, and community support activities at navy facilities worldwide. The command master chief did not want to disturb any refresher training protocols, so word was passed quietly throughout the ship. Master Chief Fa'aita intended to request the committee's consideration and endorsement to use MWR funds to pay for our support mission, one which would send six of our sailors and their four wives to Washington, DC.

Not enough sailors could make the master chief's impromptu gathering.

With Brad Ableson and Master Chief Fa'aita already on the trip manifest, we figured that we could part with no more than three to six additional crewmen without forfeiting essential skill sets and experience levels invested in most everyone aboard ship. We needed to retain specifically and critically trained personnel onboard to demonstrate to our REFTRA evaluators, as they graded us during the upcoming combat performance audit, that we were ready to deploy the *USS MARS* overseas in harm's way.

Moreover, it was not yet known how much money we were looking at spending. Airfare was the largest cost concern, especially with booking flights at the last minute with no notice. The cost of the plane tickets would be the deciding factor in determining the size of the support mission.

Normally, after getting a green light from the MWR Committee, recommendations for the expenditure of MWR monies would be brought before the MWR Council; however, only seven or eight committee members were available for that meeting.

Master Chief Fa'aita briefed me afterwards. The committee folks who did attend had championed the idea, but problematically, they represented only about one-third of the crew muster. Beyond that, it occurred to both of us that there was not time enough to call the department heads of the MWR Council together to seek a final green light.

In addition, we had a brand-new commanding officer aboard, one unknown to the crew.

Therefore, the necessity of calling the entire crew together for a meeting on the flight deck became clear. Master Chief Fa'aita and I agreed that a "Captain's Call" was in order and, with our return back to port slated for earlier than the day prior, we had a convenient time window in our day's schedule when we would be able hold one while the ship swung at anchor.

Master Chief Fa'aita and I then ran the idea past Captain Edwards, who concurred wholeheartedly.

I asked Captain Edwards that I be allowed to speak with the crew before he came back to the flight deck regarding our suggested use of MWR funds for our funeral contingent support mission, and Edwards agreed.

The "Captain's Call" was announced over the 1MC in the late afternoon.

The *USS MARS* mustered 486 men, all of whom assembled on the flight deck that afternoon, save for those on watch in engineering spaces or manning the anchor watch.

I told the men that Captain Edwards would be joining us in short order but that I wanted to speak with them beforehand on a more personal level. I gave the men a full rundown on the current thinking regarding our funeral contingent plans and the use of MWR funds. I further detailed what was already underway.

We did not have any cost estimates, but I assured the men the greatest care would be exercised in any expenditure. There would be a full and to-the-penny public accounting of payouts available on request as we progressed, and a full and detailed final accounting would be made available afterwards.

I invited questions; nonetheless, I do not recall that there were any.

I recognized that any dissenters were naturally disinclined to come forward in such a public gathering, and I said so. We had an excellent command climate wherein, as it should be, our chief petty officers ran the ship. The command master chief had an excellent rapport with his chief petty officer (CPO) Mess as well as the crew. He was always out and about the ship, and the crew knew him to be very approachable. They knew they could speak to him freely, without any apprehension. Any *MARS* sailor who was in opposition to spending MWR monies for the support mission would tell their chief as soon as we broke up from our flight deck gathering, and the chiefs knew they were free to come to me with feedback regarding the crewmember sentiments.

I suggested that sailors who did not want to approach their division CPOs could slip a written or typed note under my stateroom door or the door to the CPO Mess, leave a note in the ship's public suggestion box, or leave a note somewhere else where it could be accepted and considered.

I further explained that in using MWR monies, a fund every crewman had contributed to, that this gesture of support would include every member of the ship's company. In that moment, everyone instinctively realized this crew-wide action was a right and proper thing to do.

The men's reaction was genuinely positive, if initially subdued. Many heads were seen nodding in agreement; overall, it was "thumbs-up" all around. From my vantage point, I witnessed complete consensus approval—there was no disapproval, no upset, no grumbling, and no animosity.

Everyone agreed that this was the right thing to do, and when we were done, you could feel a burden slowly lifting from all of us. What had been a somber mood transitioned slowly into joy as it dawned on each man that there was a small, but key, role that he could proudly and personally assume to support Captain Nordeen. Thus, morale skyrocketed. The men were supremely pumped, and also ready to get on with the work of REFTRA when Captain Edwards came aft.

Command Master Chief Fa'aita escorted Mike Edwards aft to the flight deck when we were done with the support mission discussions.

Captain Edwards then addressed the troops, providing details of our situation from his perspective, and told us what he knew regarding future events and what he expected. Furthermore, he praised the men regarding their considerations in rendering respect, solace, and comfort to the Nordeen family.

Lionhearted, the crew reacted and showed they were ready to go. Their focus cleared once again and bored sharply ahead, like a laser beam.

Our chief petty officers came up with a short list of folks whom they recommended to man the support mission. Rest assured, there were a bunch of volunteers. Command master chief suggested one representative from pay grades E1-E3, one officer, and *MARS* Sailors of the Month/Quarter. Although the *MARS* contingent was the result of a tragic and somber situation, Master Chief Fa'aita thought such selection would represent the best of the crew, and further, such a nomination served as a proper and decent reward for the outstanding sailors—that is, provided their experience and skill sets were not needed for REFTRA.

Trip planning had ramped up, we had some cost figures roughed out, and finally, we determined that we could send five men picked off the chief's list.

I gave Captain Edwards an update. I requested permission to send the newly identified additional five men, along with Lieutenant Ableson and Master Chief Fa'aita, on the funeral party contingent. I assured our new commanding officer that there would be no degradation in operational performance by allowing these six crewmen to attend to their temporary off-ship military assignment.

Additionally, I explained that we had the crew's full backing in the use of MWR funds and ran down the expense numbers as we had them penciled out. For the entire group of eleven people, seven sailors and four navy wives, we had calculated a budget of approximately $11,000.

Captain Mike Edwards, in his capacities as both the commanding officer and MWR fund administrator, granted his permission for us to proceed.

Brad, Poasa, and the four crewmen who left the *USS MARS* on their journey in the late afternoon first discussed the concept of sending a team of people to Greece with Master Chief Fa'aita. When the two men came to me, the idea that our command should send a contingent of shipboard personnel in support of our CO quickly spread through the *MARS*.

In piecing together the group of people to include in our support party, it was instinctual that we must include navy wives. We were aware that of the four women in the Nordeen family, two of them were navy wives. There was also Mike and Bill's mother, Edna, and William's twelve-year-old daughter and only child, Annabel.

It was not known at the time that the two brothers also had one sister, Gene.

Thus, four women came to mind right away who could stand with the four Nordeen family members as escorts.

My first reaction was to agree completely. So, the discussion necessarily turned to obvious questions regarding how to pay for this trip and what military funding source might be available. And before we went outside the ship's lifelines seeking assistance, elementary planning called for us to decide how many to send out of the very small group of sailors whom we could afford to send. The *MARS* was, after all, just beginning refresher training (REFTRA) that was nearly an all-hands-on deck proposition throughout the entire training period.

Adding to early considerations was the element of time. Simply stated, we did not have any. Whatever it was we might be able to accomplish, we needed to do it RIGHT NOW!

And so it came to pass that we found a way to send a funeral contingent party of ten people.

Hospital Corpsman Collins had cogitated on our obvious logistical impediments and thought that the best way to proceed was to use monies from the *USS MARS* MWR fund. Fortunately, Petty Officer Collins was his Medial Department's representative on the *USS MARS* Morale,

Welfare, and Recreation Fund Committee.

Petty Officer Collins had nailed it. His was the right answer—indeed, it was our only answer!

Why? Central to the military virtue and culture in 1988 was a deep abiding respect for the military spouse.

The Fleet Training Group had granted its permission for the entire *MARS* crew to enjoy liberty over the approaching Fourth of July weekend. My pregnant wife, Cathleen (Cathy), and my kids had driven down from Oakland. Chaplain Ableson's wife, Julia, rode along with them, making the long drive south. I was able to go ashore to meet with the two women parked at the boat landing that Wednesday evening.

Master Chief Fa'aita and Lieutenant Tim Archer, our ship's medical officer, were asked to call their wives, Stella and Judy respectively, to see if they could make the trip east.

Master Chief Fa'aita, Doctor Archer, Chaplain Ableson, and I had gathered and each decided, should our spouses agree, to send our wives at our own personal expense. Moreover, it was agreed that the command master chief and chaplain would accompany their spouses, an expenditure which would also be out of pocket.

Obviously, Dr. Archer and I could not make the trip east to Greece due to our shipboard duties, as we faced the real-world operational demands of refresher training.

Cathy and Julia had heard the news on the radio regarding Bill Nordeen's assassination while driving south. They had anticipated and were prepared to receive our request before we even thought to advance it to them.

Mike Edwards had preceded me to the beach in on one of the FTG boats. Edwards needed to make some personal calls, and after that, he had to call to Rear Admiral Toney in San Francisco and others to update them on the rapidly evolving events aboard the *USS MARS*. Please remember, dear reader, that there were no cell phones back then!

When Mike Edwards returned to the boat landing, he approached me as I stood next to our Toyota truck talking to Cathy and Julia.

I walked with Edwards down the length of the pier toward the boat landing where he caught a ride back to the ship. My eleven-year-old son tagged along, listening eagerly as Mike Edwards and I walked and talked our way down the pier. No doubt, my son wanted to know what nasty circumstance had made a mess of the family's mini vacation to the southland. Our short walk allowed me enough time to run through the sketchy preliminary plans evolving at that time.

Cathy, my kids, and Julia left right from the dock to drive back to Oakland late Wednesday afternoon. I vividly remember my two boys' bitter disappointment in missing out on what had been planned for all of us: a trip to Sea World.

4.
DOMINION OVER THE ANIMALS

"It had been felt that the existence of a farm owned and operated by pigs was somehow abnormal and was liable to have an unsettling effect in the neighborhood [...]. All the year the animals worked liked slaves. But they were happy in their work; they grudged no effort or sacrifice [...]."

—George Orwell, *Animal Farm* (1946), Chapter VI

"A country's character is measured by the way it treats its veterans."

—Unknown

"They were all slain on the spot. And so the tale of confessions and executions went on, until there was a pile of corpses lying before Napoleon's feet and the air was heavy with the smell of blood [...]."

—George Orwell, *Animal Farm* (1946), Chapter VII

⚓

A CENTRAL IDEA ADVANCED IN this work is that America's military discipline system is extra-constitutional (unconstitutional, or outside the constitution) and extralegal (outside the law), and that it is purely a function of command wherein military commanders practice absolutism,[13] giving them a free hand to treat their underlings as "so many chickens."

If a given military rule bears any resemblance to the scope and

---

13 As practiced, for instance, by French King and military commander Louie XIV: "L'Etat, c'est moi."

operation of the US Constitution, it is simply a matter of that commander's passing whim.

John Jay told George Washington this much: "Let Congress legislate. Let others execute. Let others judge."[14] The bedrock foundation of the Constitution was a separation of powers in times of peace and tranquility. Little known, though, and terribly unappreciated is that the founders further provided that in times of national emergency or dire threat, all power must migrate to and reside in an "energetic executive."[15] In other words, the Constitution is situational, and its structure and power are tied directly to particular levels of threat that the country may face.

If follows that those not bound by constraints of the Constitution and those who deny constitutional protections to others cannot be trusted to defend the Constitution. Individuals and groups who deny these protections are most certainly suspect and represent a greater danger to the country than America has ever faced.

⚓

"[...] and what did he find? Not only the most up-to-date methods but a discipline and an orderliness which should be an example to all farmers everywhere. No question, now, what had happened to the faces of the pigs. The creatures outside looked from pig to man, and from man to pig, and from pig to man again; but already it was impossible to say which was which."

—George Orwell, *Animal Farm* (1946), Chapter X

14 Kohn, Richard, H. (1991). The United States Military under the Constitution of the United States, 1789—1989. New York: New York University Press, p. 43, n. 6 as follows: John Jay to George Washington, 7 January 1787, in Henry P. Johnston, ed., The Correspondence and Public Papers of John Jay, 4 vols. (New York, 1890—1893), 3:226—29.
15 Kohn, Richard, H. (1991). *The United States Military under the Constitution of the United States, 1789—1989*. New York: New York University Press, pp. 43, 48, and 80.

Offered below are a few insights for those who might wish to rail against the idea that military men today are treated like animals.

The ancient horror of military discipline, in its purposeful design to break, tame, and train civilians to become military men, is modeled after the proven success of using brute force and fear tactics with animals. Thus, it is suited to a lower class of people not deserving of a higher standard of respect.

The need of any culture or nation to maintain good order and discipline over its security force or military organization predates the need to write down those methods, particularly those which might constrain a king or commander.

Please consider, for instance, this example from the land forces: Sun Tzu seems like a very cruel man. He killed a lot of people with the troops he commanded. To those that understand the deeper lessons of his work, however, a great deal may be learned about how to reach any objective by employing a close study of what he taught.

One of the best, and I believe one of the most fundamental, lessons of Sun Tzu is that military discipline is a function of command.

Chinese King Helü (514–496 BC) wanted a demonstration of Sun Tzu's theories in action to measure their effectiveness.

To meet the king's challenge, Sun Tzu assembled the king's three hundred concubines. He organized the women into two companies, assigning each of the king's two favored women to command the others. Sun Tzu gave the women armor and weapons. Then he explained and detailed a set of drills Sun Tzu wanted the women to conduct.

Sun Tzu meticulously showed the female commanders the drill maneuvers he wanted the subordinate personnel to conduct. Sun Tzu then repeated his orders to the women commanders, directing them to lead their companies in performing the drill sequence.

The women were not warriors.

They did not believe Sun Tzu to be serious; they laughed at his directions.

Sun Tzu repeated the orders.

A second time, the women commanders did not obey and giggled in ridicule some more.

This is when Sun Tzu spoke these words: "If the instructions are not clear, if the orders are not obeyed, it is the fault of the general. But if the instructions are clear and the soldiers still do not obey, it is the fault of their officers."

Sun Tzu then ordered the female leaders beheaded.

While shock washed over the king, Sun Tzu placed two other women in command.

Sun Tzu issued the same orders regarding the drill maneuvers the women were to conduct. This time, the surviving leaders carried out Sun Tzu's exercise flawlessly.

Sun Tzu's orders had been clear each time.

The fault was with those who had been entrusted to carry out those commands.[16]

Once that problem was remedied, everything worked as intended.

Two oceanic examples that emerged onboard ships from the ancient sea forces in the practice of ensuring good order and discipline were keel-hauling and flogging.

⚓

An ancient military rule, one which is still pertinent today, informs us that any activity corrosive to the maintenance of good order and discipline represents a military crime against the state. The Roman leader Justinian

16 Tzu, Sun. (2008 illustrated reproduction): *The Art of War: The Complete and Fully Illus-trated Edition of Sun Tzu's Philosophical Masterpiece with Four Other Classics in the Ancient Warrior Tradition*. China: Sweetwater Press, p. 33.

recorded, "Every disorder to the prejudice of common discipline, such as laziness, insolence or idleness, is a military offense."[17] Included in the logic of military discipline is that a troop never knows what—in the mind of the commander—is or is not a crime against military virtue.[18]

The practice of treating a country's servicemen like farm animals has a deep ancestry that can be traced back to Roman-Britain parentage, one sharing the same ancestry with America's system of military discipline. In the context of this work, it is central to understand that the procedure historically used to maintain good order and discipline over the lower animals has been, and remains, the court trial.

In his book *The Trial: A History from Socrates to O. J. Simpson* (2005), the Finnish-Pakistani lawyer Sakadat Kadri writes: "Beginning at a time when bishops, lords, and kings were tussling over territory and peaking during Europe's sixteenth-century religious schizophrenia, they reflected the extent to which the right to judge served to express power. A mastery of the animal world was an easily understood symbol of dominion—as it had been since Rome's emperors would slaughter brutes by the thousands, just to show that they could—and the infliction of pain on beasts had an even more sinister aura."[19]

Holding comedy courts to prosecute make-believe crimes as an expression of dominion and power over animals is no different than prosecution by make-believe judges, juries, and lawyers in make-believe trials expressing power and dominion over real people in service under America's modern armed force.

Suffering under an outdated British system of attainder, pains, and penalties, America's armed forces are no better off now than were Orwell's "beasts of burden" in England.

---

17 C.E. Brand (1968). *Roman Military Law*. Austin, TX: University of Texas Press, p. 110, and Lowry, Thomas P. (1997), Tarnished Eagles: The Courts-Martial of Fifty Union Colonels and Lieutenant Colonels, Mechanicsburg, PA: Stackpole Books, p. 3.

18 Articles 133 and 134 of America's Articles of War (the UCMJ): **http://www.au.af.mil/au/ awc/awcgate/ucmj.htm**

19 Kadri, Sadakat. (2005). *The Trial: A History from Socrates to O.J. Simpson*. New York: Random House, p. 159.

⚓

Like America, Rome had originally been founded by force of arms. In Rome, the new king Caesar Augustus then prepared to give the community a second beginning, this time on the solid basis of law and religious observance. These lessons, however, could never be learned while his people were constantly fighting war. Augustus well knew that war was no civilizing influence and the proud spirit of his people could be tamed only if they learned to lay down their swords. Accordingly, at the foot of the Argiletum—a street in the ancient city which crossed the popular district of Suburra up to the Roman forum—Augustus built the temple of Janus to serve as a visible sign of the alternations of peace and war. Open, it was to signify that the city was in arms; closed, that war against neighboring peoples had been brought to a successful conclusion.[20]

Professor Shannon French informs:

A popular household god among the pagan, polytheistic citizens of Rome was Janus, the two-faced god. Janus was thought to be an ideal guardian for the Roman home, since his double set of eyes allowed him to keep a close watch in two directions at once. But the image of Janus also provides a useful analogy for the discussion of Roman values, for in Roman culture we find vivid illustrations of two conflicting extremes: self-discipline and self-indulgence.

Now, let us roam from the Old World to the New World: Mexican General Santa Anna was not at all concerned with the lives of his men. Having suspended the Mexican Constitution, General Santa Anna annexed lands to the north, stealing away property from Mexicans and "Texans" indiscriminately.[21] Resistance to this brazen, property-grabbing military tyrant led to the ignominious attack on an old mission remem bered famously as the Alamo, a battle lasting thirteen days from February

20 Chapter 1 Endnotes Livy. (2002). *The Early History of Rome*. New York: Penguin Books, p. 52.
21 Edmondson, J.R. (2000). *The Alamo Story: From Early History to Current Conflicts*. Republic of Texas Press, p. 356.

23 to March 6, 1836.[22]

On the eve of battle, members of Santa Anna's inner circle petitioned their general to reconsider his orders for a suicidal frontal assault, arguing instead for encirclement of the outpost, sealing it off from all incoming supplies. Allowing a strategy of siege would act naturally to defeat the Alamo's defenders through gradual hunger and debilitating thirst. Further, the recommend plan suggested by senior lieutenants offered the added benefit of saving the lives of hundreds of Santa Anna's soldiers.

Officers were so stunned by the general's short and caustic reply that they remained silent, fearing a certain vengeance should they attempt any further protest.

Santa Anna remained steadfast, ordering the immediate mass attack on the Texican outpost, regarding the lives of his men as important only as the loss of "so many chickens."[23]

Moving to more modern times, German-born Henry Kissinger held the sanctity of life regarding American military personnel no more valuable. While serving as President Nixon's national security advisor, Kissinger dismissed American troops as nonfunctional, except as "dumb, stupid animals to be used" for foreign policy.[24] Kissinger's disparaging remark was made, suggesting a specially intended belittling effect, in the presence of his military aide, then Army Brigadier General Alexander Haig.

Using the same condescending attitude, Don Hewitt, CBS *60 Minutes* executive producer in the eighties, called *60 Minutes* correspondent Monika Jensen-Stevenson to spike an investigative segment she was working on about forgotten Vietnam prisoners of war (POWs). During the call, Hewitt described two Green Beret soldiers as "fanatics," "crazies," and "Rambos."

---

22 Edmondson, J.R. (2000). *The Alamo Story*

23 Edmondson, J.R. (2000). *The Alamo Story*.

24 Bernstein, Carl, & Woodward Bob. (1976). *The Final Days*. New York: Simon Schuster Paperback, p. 194.

"You're worse than Kissinger," Stevenson replied, trying to salvage her television project.

"What?" said Hewitt.

Jensen-Stevenson then scolded Hewitt for disrespecting the two servicemen. She informed her boss of the 1976 Woodard and Bernstein book *Final Days*, quoting Kissinger, saying, "Military men are dumb stupid animals to be used as pawns for foreign policy."[25]

Kissinger's condescending position regarding the existence and condition of American POWs is well-known among the military. Putting action to word and thought, during negotiations to end the Vietnam conflict, Kissinger considered the existence of American POWs in Cambodia as merely inconvenient and troublesome. Therefore, during those talks, Secretary of State Kissinger denied the slim possibility that soldiers might still be alive in Southeast Asia, incarcerated in dismal settings.[26] In the end, Kissinger's settled "foreign policy" proved every bit as effective and dismissive a death sentence as Santa Anna's 136 years earlier.

Once, during the Revolutionary War, Alexander Hamilton advanced the concept to His Excellency George Washington, and to the president of the Continental Congress, that black slaves should be recruited and enlisted into service. In that letter, a researcher may discover a sincere depiction of what Hamilton, Washington, John Adams, and like-minded men thought represented the perfect soldier:

> I have not the least doubt, that Negroes will make very excellent soldiers [...]. I frequently hear it objected to the scheme of embodying Negroes that they are too stupid to make soldiers. This is so far from appearing to me a valid objection that I think their want of cultivation (for their natural faculties are probably as good as ours) joined to that habit of subordination which they acquire

---

25 Monika Jensen-Stevenson, & Stevenson, William. (1990). *Kiss the Boys Goodbye: How the United States Betrayed its Own POWs in Vietnam*. Dutton Books, p. 97.
26 *Kiss the Boys Goodbye*.

from a life of servitude, will make them sooner (become) soldiers than our White inhabitants [...]. An essential part of the plan is to give them their freedom with their muskets. This will secure their fidelity, animate their courage, and I believe will have a good influence upon those who remain [...].[27]

Esteemed mystery writer Dashiell Hammett (1894–1961), in his novel *The Dain Curse* (published by Alfred A. Knopf in 1929), expresses a similar sentiment in writing, "He was the perfect soldier: he went where you sent him, stayed where you put him, and had no idea of his own to keep him from doing exactly what you told him."

George Orwell's *Animal Farm* character Boxer, who served first under a commander named Major, then another named Napoleon, is the epitome of a perfect soldier. Boxer exhibits the ideal combination of stupidity, want of cultivation, and "natural faculties" such as strength, integrity, determination, and leadership. Of course, all of these inherent talents and qualities corrosively, and unceasingly, can be exploited by a cruel commander to a predictable end.

At this juncture, it is appropriate to mention that pre-Revolutionary recruiting inducements for slaves offered "freedom" or the award of retaining one's musket at the end of the conflict. Those older pledges are replaced in the modern era by promises of citizenship for non-citizen enlistees (e.g., the Department of Defense's guest worker program), free college education for those unable to afford a college program unassisted, or maybe just a job for the unemployed or "*new glory*" for others.[28] Today, lavish monetary bonuses and extravagant finder's fees abound, especially

---

27 Wiencek, Henry. (2003). *An Imperfect God: George Washington, His Slaves, and the Creation of America.* New York: Farrar, Straus, and Giroux, p. 226. John Laurens, Hamilton's close comrade, was first to promote the idea of recruiting a "regiment of blacks" to avert what Laurens considered an "impending Calamity," due to the shortage of freemen recruits.
28 Peters, Ralph. (2005). *New Glory: Expanding America's Global Supremacy.* New York: Penguin Group. Expound on Peters book.

in those hard-to-fill military occupational specialties (MOS).[29]

The law of the legions engendered a taming of the lower classes when early Roman armies and navies became patrician forces led by king-generals[30] who, in turn, were assisted by their retinue, including the judge marshals or judge advocates.

In the Ancient World, the best-known and most enduring codification of military law was the *magistri militum*, the law of the legions of Imperial Rome.[31] This code, which doubtless drew on customs and traditions as old as warfare itself, recognized distinctions between civil and military law that have endured to our own time. Roman military discipline was based on the simple proposition that soldiers should fear their own officers more than the enemy.[32]

The centurions were authorized to chastise with blows, the generals had a right to punish with death; and it was an inflexible maxim of Roman discipline, that a good soldier should dread his officers far more than the enemy.[33]

29 Herring, Hubert B. (January 2, 2005). *Battle Pay is Better When Wall Street is the Battlefield. New York Times*, p. A2 Bernstein, Nina. (January 15, 2005). Fighting for U. S., and for Green Cards. *New York Times*, p. A14; (June 26, 2005). To Help With Recruiting Guard Enlists Temp Service. Associated Press in *The Kitsap Sun*, p. A9. Alvarez, Lizette. (February 9, 2006). Army Effort to Enlist Hispanics Draws Recruits, and Criticism, *The New York Times*, p. A1; Cave, Damien. (2005, November 18). Vital Military Jobs Go Unfilled, Study Says. *The New York Times*, p. A16; Cave, Damien. (2006, February 5). For a General, a Tough Mission: Building the Army. *The New York Times*, p. A18; Vanden Brook, Tom. (2005, December 2–4). Guard entices ranks to recruit. *USA Today*, p. 1A; Sedensky, Matt. (2005, November 27). *To Escape... To Transform... For Money.* Associated Press appearing in *The Kitsap Sun*, p. A3; Biskupic, Joan. (2006, March 7). Court upholds military recruiting law. *USA Today*, p. 3A.
30 Adkins, Lesley & Adkins, Roy (1994). *Handbook to Life in Ancient Rome.* New York: Oxford University Press, p. 51.
31 Valle, James E. (1996). *Rocks & Shoals: Naval Discipline in the Age of Fighting Sail.* Annapolis, MD: Naval Institute Press, p. 29. First published in 1980.
32 Valle, *Rocks & Shoals*, p. 29.
33 Gibbon, Edward. (2000). *The History of the Decline and Fall of the Roman Empire: Abridged Edition.* (David Womersley, Ed), p. 18, and Valle, p. 29.

It is my belief, having lived through it, that the relationship between soldiers, sailors, and their government has never been healthy—that is, healthy for *both* sides of the equation. This is key! In America, even today the corrosive situation remains particularly manipulative. Recognizing the lower military class as ordinary farm animals, beasts of burden good for little else except to serve as government sacrifice, is as common a metaphor as it is descriptive, especially considering that a large population of soldiers were once farmers, if not slaves.[34]

Australian novelist Steven Pressfield (b. 1943) wrote *Gates of Fire*, a work of historical fiction detailing the battle of Thermopylae (published by Doubleday in 1998). Therein, he describes a side of the military community where "not everything unseen is noble."[35] Further, he writes how "It would be unnatural for base emotions of fear, and greed, and lust to totally absent themselves from military plantations. Esteem and unquestioned integrity attaching to soldiers trusted by other soldiers under fire comes from a pragmatic understanding of how easy it is for every man to cut and run."[36]

It is a fact that while senior officers enjoy an unfettered capacity to weed out subordinates who might place other lives at risk, on the other hand, subordinates have no safe way of challenging the perhaps flawed character of their commanders. Worthy of our time and attention, I suggest, is the duplicitous and corrosive nature of some—not all—military monarchs. How and when does a soldier or sailor know precisely which face of Janus his commanding officer wears? Additionally, how does someone in the lower ranks address a commanding officer who instills only fear through threats and unnecessary punishment? Obviously and conversely, as Hermann Wouk's novel *The Caine Mutiny* (published by Doubleday in 1951) tells us, proper leadership eliminates fear by

---

34 Gibbon, Edward. (2000). *The History of the Decline and Fall of the Roman Empire: Abridged Edition.* (David Womersley, Ed), p. 18, and Valle, p. 29.
35 Pressfield, *Gates of Fire*, p. 379.
36 Pressfield, *Gates of Fire*, p. 379.

consistent demonstrations of physical and moral courage.[37]

The abuse of America's armed forces under an ancient form of Roman-Britain military government and discipline continues today as if those servicemen and servicewomen were nothing but lowly farm animals. Frankly, what we have inherited—and, unfortunately, what we continue to enforce and apply—is the modern-day operation of George Orwell's *Animal Farm*; the comparisons are as irresistible as they are obvious. These pages explain in accessible English how Britain's system came to America and how it presently functions. One may study how senior military governors consistently are able to eliminate challenges to their abuse of authority and how venal politics plays a salient role, either by putting soldiers in military prisons or by inflicting other severe, life-altering punishments.

In what we would consider ancient times, members of the lower class were less educated and less refined compared to nobility. Yet today, the difference between nobles and commoners, the leaders and the led, has remained fixed, thus describing the current difference between officers and enlisted. It follows that an unfair warriors' code would naturally evolve, one responsive to the need of leadership to extract a severe level of discipline between the two basic class societies.

Thus, over the decades, instead of enjoying standard constitutional rights, servicemen have become mere property or chattel, second-class citizens.

Their lives, liberties, and pursuit of happiness are rendered insignificant and by all accounts worthless, particularly when measured against the more important issue: the survival of the state.

Therefore, over time it was determined that standard jury trials were unworkable and had to be eliminated. Judges and attorneys corrosive to good discipline were misplaced, extraneous, and had to be replaced with the arbitrary indulgences of a stern disciplinarian: the commander, His Excellency.

---

37 Herman Wouk's Second World War novel, *The Caine Mutiny*, for instance.

Over many years it was deemed by those in charge of the system of military discipline that any judicial oversight, any prudent surveillance (one perhaps involving civilian members) would be too obtrusive, invasive, cumbersome, inefficient, and obstructive. All cautionary supervision had to be neutralized. All outside interference must be eliminated, lest it compromise the safety of the state.

Cries for remedy and relief stemming from complaints of abuse will be indulged by the very commanders inflicting the indecencies, rather than examined by independent juries under law. Decisions regarding punishment must be the exclusive province of supreme commanders. Military despots populating the command structure will be tolerated for the greater good of the nation-state. Tyrants will be excused at the expense of the troops. The majority of officers decided, over many scores of years, that this was the only scheme that would work.

Monarchs, Excellencies, and senior commanding officers must never be diminished in the eyes of subordinates. The mantra was "Always above reproach"—they themselves must never be seriously punished. Excellencies must be immunized, protected from the very system of discipline that they, in turn, will turn around and inflict upon their subordinates.

It was understood that military people cannot be citizens and soldiers at the same time. *The troops will understand. We will make it up to them. And we will make it right for their families.*

The "Newburg conspiracy" was (and remains) instructive for federal lawmakers and military commanders by way of illustrating just how much the government could (and can) get away with in its abuse and exploitation of servicemen faithful to their country. Succinctly, America came closer to a military overthrow of civilian government than ever before in its history. His Excellency George Washington quieted the percolating revolt, and by so doing, perhaps inadvertently, set the foundation regarding the high level of subservience Americans expected of their military.

Thus, unsurprisingly, all three branches of US government, along with handpicked senior military governors, have exploited their many powerful advantages over the lower military classes ever since. The Bonus Army, briefly discussed later, serves as a salient example amongst many.

Orwell's *Animal Farm* metaphor serves its greatest purpose by demonstrating exactly how clever and silent assimilation, or achievement of overarching power, is accomplished. In his persistent and brilliant artistry, Orwell illustrates through logical and lyrical sleights of hand that war is peace, freedom is slavery, and ignorance is strength. In their manipulative writings on military discipline, Winthrop and his princelings, for their part, are interested in having the entire civilian community believe, through similar acts of clever and silent assimilation, that courts-martial are not only the same as civilian courts but, indeed, fairer and more impartial. What a clever sleight of hand Winthrop has accomplished!

Accordingly, Winthrop and his apostles want the general citizenry to believe military judges are the same as federal judges—that there are appeals where none exist, and that "military due process" is perhaps even better than constitutional due process. These people are responsible for renaming the Articles of War to the Uniform Code of Military Justice (UCMJ), and further, with extreme farfetchedness, they celebrate the UCMJ as a government organization superior to that found in the Constitution.

Military officers trained in the law, judge advocate generals, and the commanders they serve over time have become quite practiced in advancing this false comparison between the court-martial discipline system and civilian trials. In truth, there is no equivalency! Indeed, no two schemes could be more dissimilar; they are in no way comparable except by way of fancy linguistic trickery. Any close study of Winthrop's work (and others') offers his calculated deception that military disciplinarians function under orders in exactly the same manner as civilian judges and juries do (those who operate independently and fairly in their roles and administrative duties), and there is no statement on military law that could be more duplicitous and treacherous.

5.
THE ULTIMATE BETRAYAL

"The power of the lawyer is in the uncertainty of the law."
—Jeremy Bentham (1748–1832), English philosopher

"Nobody likes a JAG."
—The JAG Hunter

ONE MUST BE TOLD SOMETHING if the gist is unclear. Or, one must look it up.

Some readers may be aware that the above depiction is the insignia identifying naval personnel, one whose military occupational specialty is military discipline. But what, precisely, does one observe?

The mill rind, depicted on the uniform insignia worn by navy attorneys, signifies that they are members of the Judge Advocate General Corps. The "rind" is part of a machine used, for example, to grind wheat into flour. The rind connects the uppermost grinding stone to a turning spindle.

Two oak leaves embrace the machine part. Here, oak is used as a symbol for strength—or, think, *power*.

⚓

In the motion picture adaptation of *The Americanization of Emily* (the iconic screenplay was written by Paddy Chayefsky in 1964), Admiral William Jessup, excellently portrayed by Melvyn Douglas, carries out the greatest sin that a senior officer can commit against a subordinate: He knowingly sends troops to their death for personal motives or gain. In the riveting picture, a crazed and demented navy rear admiral plots to ensure that one of his own staff officers, a sailor, becomes the first man to die on Omaha Beach during the impending June 6, 1944, amphibious assault, the Normandy invasion we know as D-Day. A fellow staff officer, an Annapolis graduate, is so obsessed with the conduct and success of the venture that he directs his staff colleague to force the fore-condemned officer to advance up the beachhead when the offensive is launched.[38]

⚓

"Who smote Abimelech the son of Jerubbesheth? did not a woman cast a piece of a millstone upon him from the wall, that he died in Thebez? why went ye thy nigh the wall? then say thou, Thy servant Uriah the Hittite is dead also."

—2 Samuel 11:21 (King James Version)

"But the thing David that had done displeased the Lord."

—2 Samuel 11:27 (KJV)

King David's sin with Bathsheba is well-known, but Commanding General David's order ensuring the certain slaughter of a subordinate

---

38 *The Americanization of Emily*, Arthur Hiller, James Garner, Julie Andrews, James Coburn, Edward Binns, Keenan Wynn, Warner Home Video, Inc., 2005. Original release date 1964. Based on the novel by William Bradford Buie.

soldier, Uriah the Hittite, stands out in military culture as David's greater sin and a supreme act of supremacy, a sin probably less understood to those not interested in complicated matters of military discipline.

David's assassination of Uriah stands out as an expression of the ultimate betrayal of his soldiers and as a supreme act of supremacy. Therefore, as previously mentioned, it is the transcendent military crime that a commanding officer can commit against a subordinate.

One of the weapons used in the battle at Thebez, thrown down from a wall upon the troops standing with Uriah, was a broken piece of a millstone.

This work, Clandestine Acts of Supremacy, is a painstaking step-by-step modern-day recording that demonstrates how a senior military officer in command, one Rear Admiral John Bitoff, utilized the JAG's great grindstone to destroy a subordinate, all done by employing a clandestine act of supremacy.[39]

⚓

John Bitoff court-martialed me for "stealing" the money used to send the *USS MARS* contingent funeral party to Captain Bill Nordeen's military funeral in Greece. Years later, Bitoff admitted:

"I brought the charges, and I convened the court-martial."[40]

I firmly hold that Bitoff's revenge upon me was an outright malicious act of calculated revenge. His vindictive behavior displayed a deep-seated arrogance, one characteristic of those obsessed by an unabashed avarice for rank and of those who intend to destroy anyone who might stand up to them or call them to true account. Therefore, this work is a real-life recording of what actually took place that recounts how a senior military officer in command, John Bitoff, utilized the JAG's great grinding stone in an Act of Supremacy.

---

39 The reference here is to Henry VIII's notorious Acts of Supremacy.
40 Appendix One: Bitoff letter of 30 April 1999.

Unfortunately, no judicial review was ever to take place in my future. Instead, in the following note of May 4, 1998, from Leon Carroll Junior to all Puget Sound personnel, please read and study how I was consistently blackballed and placed off-limits, thus made a persona non grata in my own beloved navy:

SPECIAL AGENT IN CHARGE NCIS PACIFIC **b7C** NORTHWEST

```
From:     LEON CARROLL, JR.
To:       W.\OFFICE31\GRP\PSFO.GRP
Date:     Monday, May 4, 1998  1:23 pm
Subject:  OFF LIMITS
```

This is a notice to all Puget Sound personnel:

A Mr. Walter Fitzpatrick has made inquiries regarding an event that happened to him several years ago when he was the XO of the USS Mars homeported in San Francisco. He was administratively discharged from the navy and is know claiming he was framed. While residing in the Washington, DC area, he made a complaint to our DC office that a memo with his signature forged was used in the proceedings. DCWA opened a case and had the handwriting examined and the results were inconclusive. There is nothing more we can do for Mr. Fitzpatrick. He has now levied charges against his former defense attorney, now a deputy prosecutor with Kitsap County alleging that it was the attorney who forged his signature. Again this has been investigated and the case closed by DCWA. **b7C**

This morning Mr. Fitzpatrick arrived at the Bremerton Office to file the same complaint. Fortuunately RAC MARY CALL was familiar with the situation and explained to Mr. Fitzpatrick that NCIS had looked into his complaint and could do nothing further to help him. He departed NCISRA Bremerton stating he was going to take his case to U.S. Rep. Norm Dicks.

Mr. Fitzpatrick has shopped his story around for years and has reached the point of shere deparation. He is not to be allowed access to any of our spaces. If he shows up or calls your office politely tell him that NCIS has looked into to his complaint and any further information regarding his case can be obtained by quering our headquarters through the Freedom of Information Act.

b7C

CC:

RECEIVED

TUESDAY, 11 AUGUST 1998

NCIS FOIA RESPONSE

Note: I lost my sword and its case and knot. If anyone knows where it is contact the publisher. My name is on the sword blade.

6.
ANATOMY OF A COURT-MARTIAL: ZELLER'S MEMOS

"I can think of no more fitting expression of this country's appreciation for the sacrifice of our young servicemen than to grant them the same rights they are defending."

—Senator Sam James Ervin Jr. (1896–1985)

⚓

WHAT FOLLOWS IS A STEP-BY-STEP depiction of my court-martial at the behest of the United States Navy. At times, it is painful for me to go over the events in my mind, to essentially relive the details of that betrayal; however, perhaps all of us may learn something from that further study. Accordingly, here goes: On October 4, 1989, Rear Admiral John Bitoff first received permission for an investigation from a three-star vice admiral named David M. Bennett (Commander, Naval Surface Force, Pacific).

After that permission was secured, the horses were off to the races!

Bitoff took charge. Contrary to the rules, Bitoff then acted as both my senior investigation officer and prosecutor. As the convening authority, he writes: "I brought the charges, and I convened the court-martial."[41]

Following that action, he dismissed the Naval Investigative Service (as it was known in those days before it had to change its name to the Naval Criminal Investigative Service) from any inquiry. In doing so, Bitoff eliminated any possibility of outside investigation.

---

41 Appendix One: Bitoff letter of 30 April 1999.

DEPARTMENT OF THE NAVY
COMMANDER COMBAT LOGISTICS GROUP ONE
FPO SAN FRANCISCO 96601-5200

IN REPLY REFER TO

5041
N14/1322
4 Oct 89

received
TUES 31 DEC 91

FOR OFFICIAL USE ONLY

From: Commander, Combat Logistics Group 1
To: Commander, Naval Surface Force, Pacific Fleet

Subj: HOTLINE PROGRESS REPORT

Encl: (1) HOTLINE PROGRESS REPORT ON 890825
        (2) HOTLINE PROGRESS REPORT ON 890863
        (3) HOTLINE PROGRESS REPORT ON ⬛⬛⬛⬛⬛⬛⬛

1.   Enclosures (1) through (3) are forwarded for your
information.

P. A. ROMANSKI
By direction

FOR OFFICIAL USE ONLY

ENCLOSURE (5

Bitoff then pronounced me guilty of a crime. For further details, please see and study the Zeller investigation report.

Next, he assigned his own attorney to investigate, then Lieutenant Timothy W. Zeller. Remarkably, this action would put a junior officer in a position to investigate a senior. See Lieutenant Zeller's ominous

Thanksgiving Day message: "We picked the defense counsel, Lieutenant Anderson, so that the hearing would be fair."

Next, John Bitoff ordered his assigned defense attorney to my case, Marine Lieutenant Kevin Anderson, an action clearly counter to proper justice.

Bitoff then rigged the Article 32 hearing, a preliminary hearing which is required before referral to a general court-martial, by threatening a former subordinate.

Following that action, and while remaining ever-alert and manipulative, Admiral Bitoff rigged the court-martial, inserting his own staff. In particular, he inserted one person who had been previously removed due to his prior conflicts with myself: one Steve Letchworth.

After the court-martial, Bitoff used a phantom witness, Lieutenant Commander Doug Dolan, to testify against me. Dolan was not interviewed by Zeller during his investigation report (a salient fact withheld from the court-martial), did not appear at my Article 32 hearing, and did not testify before my court-martial.

Later, Lieutenant Anderson committed a crime by forging my name to a confession.

John Bitoff tried to extract himself from all possible culpability, writing the untruthful letter above to Representee Dicks and Navy Secretary Danzig once he knew that he was on the verge of being caught in his tracks.

Thus, planned with exquisite precision, the court-martial cover-up was launched!

At this juncture, it would be prudent to harken back for a moment to the aforementioned court-martial of Captain Charles McVay III. In that case as well as mine, one man working as both prosecutor and accuser—in this instance, Secretary of the Navy James Forrestal—acted as the convening authority against Captain McVay.

In truth, looking back at both sad events, neither case was court-martial material, nor even close to that level of gravity or seriousness. In attempting to snub supposed "insubordination," both were mock, faux, fake military trials masquerading as the real deal.

Both cases were an infraction of what today is known as the flatulent Racketeer Influenced Criminal and Corrupt Organizations Act.

Neither case was governed or influenced in any way by *Dynes v. Hoover* (1857) under US adjudication.

There was no immediate judicial oversight.

Further, there was to be no later judicial review.

It is my firm belief that all of these duplicitous and fraudulent actions could only have taken place in a godless society.

⚓

This intentional purge to extradite faithful, career-minded officers from the officer corps has been underway for a long, long time.

⚓

Lieutenant Tim Zeller writes:

In the fall of 1989 I was tasked with conducting an investigation into the MWR expenditures onboard *USS MARS*, said tasking being a result of a directive from Commander, Naval Surface Force, Pacific Fleet. My client in this matter, as both an investigator and Legal Officer, was the Department of the Navy as personified by Rear Admiral John Bitoff, the Commander, Combat Logistics Group-1.

Vindictive animus. From the outset, strong friction existed between the ship's company and CLG-1 staff.

Whenever Tim Zeller puts pen to paper in his many writings, he inadvertently reveals the narrative as to how the court-martial central to this work began and evolved. What speaks clearly throughout Zeller's writings is the information he invented and the information he decided to leave omit. At base, as the individual reader will determine, Zeller's body of work and discourse is characterized by a calculating deception.

Zeller kept most of his reports secret specifically so that Bitoff's rigged court-martial and ancillary frauds would not be ferreted out and discovered. Even today, relevant records are still being guarded as national secrets. We do not know the full extent of Zeller's tricky maneuverings since he willfully admitted to the practice of destroying documents, and many other officers joined Zeller in this continual dark pattern of concealment and destruction.

Protesting "fairness" to the accused at every opportunity, Zeller knew from the start he was not allowed to act as the investigating officer. Zeller used his professional training, knowledge, and skills to construct a military discipline obstacle course that no one could have survived. To be clear, Zeller's campaign was not about reconciling concerns respecting the operation and administration of the *USS MARS* MWR account; it was about vigorously attacking me and firmly harming my career in the navy.

In this regard, Zeller's memos offer a strong basis for a strong argument against Colonel William Winthrop and other military disciplinary professionals who have crafted our system of military justice. These military men of the past were as adamant as Zeller is today in working to persuade outside observers that the military discipline is not only "fair," but indeed a much better form of government than normal US citizens enjoy under the auspices of the US Constitution.

The uninitiated or uninformed reader may require assistance to fully appreciate the significance of Zeller's writings. Often, with cleverness and guile, Zeller uses the concept of "fairness" [*sic*] to explain his activities and reasoning. Zeller was intentionally trying to dupe the unalert or uninterested reader into thinking that all the actions that John Bitoff,

Zeller, and Anderson took in the early days of the military discipline hearing were "fair" to the accused Fitzpatrick.

However, papers so far unearthed and herein recorded betray their true purpose of deception and damage. In addition, those documents undercut and destroy the integrity and trustworthiness of the Navy JAG Corps, the Naval Criminal Investigative Service, and the military establishment as an entity.

In truth, normal fairness played no part in any of these proceedings.

⚓

Fourteen months earlier, in October 1989, Captain Edwards was one of a number of players from John Bitoff's Group ONE staff who accused his executive officer on the *MARS* of personally stealing from his shipmates those funds used to send a small contingent of people—representatives of the *USS MARS*—to Bill Nordeen's funeral in Greece as a sign of consideration and faith.

Thus, the executive officer was ordered to stand before the court-martial fourteen months down the road. It should be noted that in military circles, the act of stealing carries special censure because of the extraordinary opportunity that thieves can exploit in close-quarter army barracks or on navy ships where privacy comes at a premium. Stealing is a very serious accusation, and one not made lightly.

⚓

In September, the *USS MARS* sailed from Oakland for a six-month Persian Gulf deployment, set to return the following March. Meanwhile, Bob Toney turned over command of Combat Logistics Group ONE to Rear Admiral John W. Bitoff in January 1989.[42] Neither Mike Edwards's new boss nor a six-month separation improved the acrimonious,

---

42 Combat Logistics Group ONE has since been disbanded (decommissioned).

long-smoldering working relationship between the MARS personnel and the Group ONE staff.4 The rapid downward spiral that may describe interactions between *MARS* crewmen and Bitoff's staff only accelerated the rapid race to the court-martial.

⚓

Morale, Welfare, and Recreation programs throughout military units are funded by the troops themselves from profits resulting from base or ship stores. MWR are monies collected, banked, and spent on various expenses, as in any civilian business. MWR funds are private revenues, rendering them off-limits from the types and kinds of government oversight required for congressionally appropriated tax revenues. MWR accounts are overseen and managed by individual unit organizations that are then later reviewed by upper management.

For the *USS MARS* in the late eighties, outside MWR audits were conducted every eighteen months with scheduling responsibility resting first on the shoulders of Rear Admiral Bob Toney, then Read Admiral John Bitoff.

Annual MWR reports were studied by outside auditors in their oversight duties. MWR account reports were mailed to them no later than September 30, which is the end of each fiscal year.

As authorized by *MARS* crewmen in June/July 1988, MWR funds were used to pay for funeral expenses resulting from honors and respect rendered to Captain Bill Nordeen for his funeral in Greece. An outside audit examination of the *MARS* MWR account—to include a review of itemized expenditures found in the 1988 fiscal year report—was scheduled for June 1989; nevertheless, mysteriously, it was delayed for months by Bitoff staffers. And astoundingly, the written report and all copies, as shall be seen, somehow disappeared in late 1989.

With the ship's return from deployment, Executive Officer Fitzpatrick on the *USS MARS*, striving to calm troubled waters between ship and staff, briefed Captain Mike Edwards in the Group ONE XO's Oakland headquarters office, giving him a detailed presentation of problem areas and staff guffaws. Feedback from the briefing reached Group ONE staffers, immediately creating a "firestorm" that spread quickly and burned white-hot for months.[43]

⚓

Captain Mike Nordeen handed over the keys to the *USS MARS* on Thursday, August 31, 1989.[44]

John Bitoff ordered an MWR audit aboard the *MARS* the next day, Friday, 1 September 1989. Bitoff's scheduling of the audit was strange on two accounts. First, it was three months late, and second, it was carried out on the day after a shipboard change of command. The auditor, an MWR subject matter expert from San Diego not assigned to Group ONE, was alerted by at least one Bitoff staffer to perceived problems with the *MARS* MWR account and directed her, Ms. Ruth Christopherson, to focus intently on funeral expenditures from July 1988.[45]

In her pro forma report, Ms. Christopherson failed the *MARS* MWR account managers, pointing chiefly at funeral trip expenses as her chief complaint. Putting aside questions regarding Ms. Christopherson's independence and objectivity, the experienced MWR examiner, with good reason, did not make a single mention—by name or position—of the *MARS's* executive officer (XO) in her twelve-page audit findings. Shipboard XOs, per normal ship routine and organization, do not hold responsible positions in the management or oversight of MWR accounts.

---

43 Captain Edward's title was chief of staff, second in command to Admiral Bitoff, but his function was the same as any military executive officer: to instantly assume command when it becomes necessary.

44 The skipper was then reassignment to a new "placeholder" spot on the Pentagon's base realignment and closure commission until taking command of the aircraft carrier the *USS CONSTELLATION* (CV-64).

45 Lieutenant Bill Bramer's sworn affidavit.

Using the specific example of the *USS MARS*, it was Art Rorex—a supply officer with deep financial experience holding the rank of commander (a navy O5—in other words, a senior officer)—who was the responsible officer for immediate shipboard oversight of the MWR program. He reported directly to the commanding officer, the ship's senior MWR account manager (a navy O6—also a senior officer).

These accounting details are key if one is to fully understand what occurred after Ms. Christopherson disembarked the *USS MARS*.

⚓

Two weeks after Ruth Christopherson filed her report, higher command directed Rear Admiral John Bitoff to conduct a closer inquiry into the financial workings of the *MARS* MWR account between January 1988 and September 1989 (both dates bracket off-ship audits of the *MARS* program).

In his book published in 1956, just five years after the enactment of the Uniform Code of Military Justice (UCMJ), Professor Everett makes the case that accused servicemen now enjoy greater protections than they did in the pre-UCMJ military establishment. "The requirement of extensive [pre-hearing] investigations of serious charges before their reference to a court-martial" is one those added protections Everett profiles.[46]

However, here is how those pre-hearing investigations worked thirty years later: Enter Lieutenant Timothy W. Zeller (a navy O2), a staff judge advocate to John Bitoff and Mike Edwards.

At light speed, Rear Admiral Bitoff's first reaction was to return any investigatory responsibilities back up the chain of command. Please recall the crucial fact that it was that same Mike Edwards, Bitoff's XO, who commanded the *USS MARS* in June and July 1988 during the two

---

46 Everett, Robinson O. Military Justice In The Armed Forces Of The United States. (Harrisburg, Pennsylvania: Military Service Publishing Company, 1956), p. 2, 10. Mr. Everett was Chief Judge to the Court of Military Appeals 1980-1990.

weeks in which funding for the funeral trip was approved and monies spent. Edwards, as the ship's skipper, was routinely briefed on events as they evolved; further, he personally approved the funding source once it was identified and agreed upon, and he also okayed the crew members selected to attend the funeral who were to be absent from the ship during REFTRA while the shipmates remaining behind were to be strenuously tested.

Additionally, once those sailors were gone, Captain Edwards received a daily written record—a "muster" (attendance) report—wherein all sailors missing from the ship were identified by name, giving the commanding officer an explanation why each man was absent when reasons were known.

Apparently unconcerned by the deep conflict inherent in the admiral's intimate association with Executive Officer Edwards, Bitoff assigned his staff aviation officer, Lieutenant Commander (O-4) Steve Letchworth, to conduct the first exploration of possible financial misconduct. Bitoff was willfully ignorant as well regarding Letchworth's relationship to Edwards. Letchworth, two ranks junior to Edwards in the same organization, was placed in a situation whereupon he would scrutinize Edward's performance in command of the *MARS*, searching for evidence of alleged financial malfeasance while at the same time, back in staff headquarters, working for Edwards, who had input as to Letchworth's performance evaluation as the staff's resident pilot. Normally, this intentional and coercive conflict of interest does not lead to results that are fair to all.

Yet still, matters grew even more complex. Bitoff fired Letchworth as preliminary investigation officer before Letchworth ever began work, but not for obvious reasons.

Bitoff was advised of a near-contemporaneous professional dispute between Captain Nordeen's XO on the *MARS*, a surface warfare officer (or ship driver), and Steve Letchworth Group ONE's in-house helicopter pilot. Letchworth, directly after the *MARS*'s return from deployment in

March 1989, found what he considered a serious deficiency regarding the ship's flight deck (helicopter landing zone), and he then pressed to decertify the *MARS* for flight operations. This is a very big deal for many reasons.[47]

First and foremost, Letchworth was wrong. He was challenged regarding his determination to decertify the *MARS* flight deck and was then overruled by higher command (Bitoff's bosses) in San Diego. Therefore, the *MARS* retained her war-fighting critical certification to land and launch helicopters, to the embarrassment of Letchworth personally and the Group ONE staff more generally. The argument had run for weeks and was just beginning to simmer down a little when Bitoff tapped Letchworth as an ostensible "unbiased" finder of fact, sent to ascertain the proprietary of financial management and operation of the *USS MARS* MWR account.

It is precisely at this point that Bitoff's true intentions were laid bare and became clear, at least to staffers who Bitoff trusted to keep his secrets. Bitoff's staff judge advocate (legal advisor), Tim Zeller, reported years later about the selection process Bitoff used to replace Letchworth. As it happened, the pending inquiry had nothing to do with the *MARS* MWR account.[48]

Bitoff's genuine target was the executive officer holding the rank of lieutenant commander (O-4). Bitoff initially selected Letchworth, also a lieutenant commander, based only on Letchworth's seniority to the XO *MARS* by date of rank.[49] Once Letchworth was removed, Zeller reports Bitoff was unable to find any Group ONE staff officer who had not had

---

47 Bitoff, as a one-star rear admiral, held the seventh highest maritime commissioned officer rank—an O-7.

48 Tim Zeller's naval message to Navy Captain Glen Gonzales of 16 April 1992.

49 Seniority of one officer over another, when both hold the same rank, is determined by their relative promotion dates. For instance, an officer promoted to the rank of lieutenant commander today is senior to every other LCDR promoted the next day or later. Officers promoted to the same rank on the same day are reported on a list of names each assigned a line number. So, in this common circumstance, seniority is found through "lineal number" comparisons.

a similar "confrontation" with his target XO. He wrote, "These factors were considered to override normal protocol that a junior not investigate a senior."[50] Thus, it came to pass that Rear Admiral Bitoff selected his staff judge advocate general, Lieutenant Timothy W. Zeller, to carry out the preliminary inquiry now specifically focused on the single personage of the XO *MARS*.

I want to remind readers that Zeller was as conflicted by identical professional relationships to Edwards as Letchworth had been. Still, Bitoff did not seem to notice or much mind. Bitoff's selection, then rejection, of Letchworth and his subsequent selection of Zeller as investigator occurred in a single day, Friday, September 15, 1989.

⚓

As the *USS MARS* got underway that morning to join the Third Fleet in Pacific Exercise 1889 (PACEX '89), Zeller began his inquiry three days later, embarking aboard the *USS MARS* on Monday, September 18.

As the teller of this tale, I now feel the need to inform readers that "[in] conducting his inquiry, an investigator should be unfailing in his quest for the truth. He should remember that his job is to conduct an impartial inquiry, designed to establish all the facts in the case, and not to perfect a case whether for the government or for the accused."[51] Navy Captain and military discipline authority Edward Byrne published this job description for investigators in 1981 in the third edition of his series of works on military law. Professor Everett is in sure agreement.

It took Zeller twenty-three days to pronounce the XO *MARS* "guilty," thereby recommending to Rear Admiral Bitoff the most severe level of command focus: a general court-martial.[52] Zeller accused the XO of "stealing" monies used to send the ship's contingent to Bill Nordeen's

---

50 Tim Zeller's 16 April 1992 message to Captain Glen Gonzales.
51 Byrne, Edward. *Military Law*, p. 57.
52 Zeller draft investigation report of 10 October 1989.

funeral in July 1988, amounting to roughly $10.5 thousand.

Notable, too, were Zeller's implied accusations against the Pacific Fleet's MWR expert, Ruth Christopherson, of incompetence, if not outright criminal complicity. Christopherson and Zeller were both aboard the *MARS* conducting MWR audits in September 1989, just days apart. Christopherson's job was to examine the MWR accounts for an entire fleet of ships homeported on the West Coast of the United States, not just Zeller's.

And yet, while Christopherson did not suspect, nor did she report criminal activity of any sort (and though she did not once mention by name the XO), by contrast, Zeller's report called him out twenty-six times in a six-page narrative, accusing him of stealing from his shipmates by raiding the MWR account. Further, Zeller reported that he had come into possession of sufficient documentary evidence to declare criminal fraud at the hands of the XO, who was "guilty" on all charges and brought before a general court-martial convened by Bitoff.

⚓

We all understand that there is a vast difference between exercising bad judgment in making discretionary calls and engaging in outright lawlessness. This occurs even in the military establishment, an entity set apart from all others and one that has "effectively resisted the most significant normative forms of civilianization." Professor Mark Osiel accurately observes that the armed forces have their own "police" forces, separate entities that occasionally restrict even admirals and generals in their authority.[53] One of those agencies is the Naval Criminal Investigative Service (NCIS, formerly the Naval Investigative Service—NIS).[54]

Late in that September of 1989, Lieutenant Zeller immediately

---

53 Osiel, Mark J. (2002). *Obeying Orders*, p. 29.
54 Each military service has its own federal police force performing functions identical to each of the others. The army calls its in-house outfit the Criminal Investigative Division, or CID. In the air force, the agency is named the Office of Special Investigations, or OSI.

briefed Read Admiral Bitoff regarding his findings and judgments. Zeller told Bitoff the *MARS* XO, a commissioned officer—this writer—had pillaged the ship's MWR fund of over ten thousand dollars and had perpetrated other frauds against MARS shipmates. If anything magnifies the severity of a shipboard theft, it would be the fact that the crime had been carried out by a CO or XO or, most sinisterly, both. As it happened, and most implausibly, Zeller had also named Mike Nordeen as a co-conspirator in the robbery and fraud and found Nordeen as guilty as his XO.

The UCMJ prohibits any investigating officer later acting as a legal advisor to then act as a reviewing or convening authority on the same case. First year criminology or law students, Professor Everett, or any law professional quickly can recognize the wisdom found in the advice that staff judge advocates should "avoid conducting preliminary inquiries, if his commander has the authority to convene (or create) courts-martial."[55]

Bitoff misjudged the inquiry aboard the *USS MARS*, firstly due to the involvement of Captain Edwards in seminal events with Edwards's subsequent authority over Zeller, the assigned investigating officer, and secondly due to Zeller's job on the Group ONE staff under Bitoff and Edwards as his legal advisor. On the other hand, it was Bitoff's and Zeller's duty to relinquish all investigatory efforts to the NCIS immediately upon Zeller's serious accusations of criminal enterprise by the CO/XO of the *USS MARS*, also per Zeller's courts-martial recommendations.[56]

Zeller and Bitoff were acting as a unified team by the end of September 1989. Zeller described the association this way: "My client in this matter, as both an investigator and (legal officer) was the Department of the Navy as personified by Rear Admiral John Bitoff, then Commander, Combat

---

55 Byrne, Edward. *Military Law*, pp. 57–8.
56 Zeller's 16 April 1992 message as well as Zeller's draft investigation report of 10 November 1989.

Logistics Group ONE."[57] Trained as an attorney, Zeller knew then that the action of one man was the action for both men.

In keeping with his established attorney-client relationship, Zeller was required to work with Captains Edwards and Paul A. Romanski. As Bitoff's assistants, Edwards and Romanski also represented Zeller's clients. Zeller delivered his first briefing to Paul Romanski on October 2, 1989. Exactly two weeks after Zeller began his inquiry into what Zeller himself described as a "complex and serious situation," Zeller was prepared to deliver his initial assessment to Romanski, no doubt expressing his opinion that the executive officer (XO) on the USS Mars, Walt Fitzpatrick, as a commissioned office and second-in-command, was a shipboard thief stealing from money from the crew. Romanski, subordinate to Bitoff, and Captain Edwards ordered "Zeller to complete an initial written report (extensive) for review by Romanski and Edwards, then for presentation to RADM Bitoff."[58]

Rather than alerting NCIS operatives, Zeller prepared his first report draft and delivered it to Romanski on October 5. Together, Romanski and Edwards reviewed Zeller's report on October 10. Bitoff met with Edwards, Romanski, and Zeller on October 12. Bitoff agreed to Zeller's recommendations to include a general court-martial and ordered the requisite Article 32 examination to be convened. Next, Bitoff ordered Zeller, in the presence of Edwards and Romanski, to prefer charges against the XO MARS, this writer, with a view to general court-martial (GCM). However, Zeller's investigation was not complete, and nobody called an NCIS agent.[59]

---

57 Zeller's letter to Counselor Dan Murdock, general counsel of the Oklahoma Bar Association, of 8 July 1998.

58 Appendix Six: Captain Romanski's 3 November 1989 MEMORANDUM FOR THE RECORD, "Chronology of Events."

59 Appendix Six: Captain Romanski is author of this timeline, distributed as a MEMORANDUM FOR THE RECORD dated 3 November 1989. Zeller's October 5 draft investigation report remains secret despite unrelenting and continuing Freedom of Information Act requests. However, the October 10 version does survive, undoubtedly identical to its five-day-old predecessor.

Beyond his clear obligation to alert and energize the NCIS, Rear Admiral Bitoff also was required to notify his seniors regarding his automatic disqualification to create a contemplated court-martial. Zeller's quick finding of "guilt" for Mike Nordeen and Nordeen's XO conspicuously demonstrated—at least to close advisors who knew the case thoroughly—Zeller's and Bitoff's personal animus in a future prosecution of the *MARS* XO using only premeditated, agreed-upon allegations.

Although by law he was required to divorce himself and his command from continued involvement in the case, instead Bitoff set off on his own criminal escapade after receiving regular advice and counsel from his Group ONE legal advisor, Tim Zeller.

Zeller's investigative protocol illustrated above accurately describes the "independence" military investigators enjoy, standing in clear contrast to what Professor Everett and Everett acolytes would have nonmilitary audiences believe. Zeller's accusations were propelled into a court-martial only after two navy captain and one rear admiral vetted Zeller's report and signaled green lights. Meanwhile, outside commands, including the NCIS, were continually kept at arm's length, left completely in the dark.

Here is another fact worth mentioning: Zeller not only determined and achieved agreement from Bitoff and others regarding the need for a court-martial—he had also reached an a priori and presumptive conclusion of a verdict of guilt, even before the commencement of the trial.

For Zeller and his superiors, the formal requirement for an Article 32 determination must have been as annoying and frustrating as it was inconvenient.

⚓

In the sequence of "extensive [pre-hearing] investigations of serious charges before their reference to a [general court-martial],"[60] there arises

---

60 Everett, p. 10.

the requirement, under UCMJ Article 32, to formally examine "pre-ferred" charges along with supporting evidence.

Throughout the court-martial, Zeller did Bitoff's heavy lifting.

Professor Everett is particularly celebratory of the structural change brought about by 1951 UCMJ legislation that newly demanded assignment of competent defense counsel, a trained lawyer, to service members charged with military crimes. Everett comments that "[with] qualified counsel to represent him, the accused is shielded to a great extent against many of the other claimed injustices of [disciplinary hearings] by court-martial—such as that all sorts of improper evidence would be offered against the accused simply to build up prejudice in the [...] mind [of hearing members]."[61]

Bitoff and Judge Advocate General Corps (JAG) staff Zeller, having become formal accusers of XO *MARS* by November 1989 and in the days leading up to the Article 32 forum, were so concerned that the XO may be properly protected in an environment conducive to an unbiased hearing that they handpicked the XO's own defense attorney.

Zeller, writing in his own words in a secret internal memo to Rear Admiral Bitoff on Thanksgiving Day, made clear that "we [Zeller and Bitoff] asked for an above average counsel for the [defense] in order to ensure that the [hearing] be fair [...]."23 Neither Bitoff nor Zeller (accusers) bothered to inform XO *MARS* (the defendant), nor any outside observer, about their special interest in the selection and assignment of the defense attorney.

Preparing for the Article 32 hearing, envisioning a general court-martial and acting under John Bitoff's authority, Mike Edwards and Tim Zeller busied themselves with the collection of all documentary evidence. However—illegally, it must be noted—NCIS special agents never heard about Zeller's inquiries, nor were Zeller's follow-up investigation, results, and accusations ever made known outside Group ONE offices. Absent

---

61 Everett. P. 10.

professional consultations, NCIS agents would have provided Bitoff's staff with personnel, not to mention established investigative procedures. In other words, Zeller, Edwards, and others were free to gather all and any MWR records without cumbersome chains-of-custody strictures or itemized listings of collected papers.

"In order to ensure" that the roles filled by Bitoff and Zeller were "apparent," Bitoff directed Zeller to sign charge sheets for them both as accusers against XO *MARS*.[62]

As a further safeguard and for greater expediency, Bitoff thought it prudent that all physical evidence be held in the custody of his co-accuser, Zeller. And so it happened that when any evidence came into his possession, Zeller immediately locked it in his safe behind his bolted office door in the Group ONE headquarters building, which is also locked and patrolled by base security onboard the Naval Supply Center in Oakland where guards were posted at every open gate. Access to Zeller's inner sanctum was very tightly controlled.

Advocating for an aggressive government approach,[63] Zeller and Bitoff not only selected Marine Captain Kevin M. "Andy" Anderson (a marine corps O-2) as defense counsel for XO *MARS*, but the two men also took an active hand in selecting the presiding officer to the Article 32 hearing. And while it is true that Zeller and Bitoff "did not ask for a certain GC [government counsel, or prosecutor] [...] due to the command influence factor," once Zeller recognized that Navy Lieutenant Matthew Bogoshian had failed to follow orders by "repeatedly refusing to repeat [Zeller and Bitoff's] position on witnesses, and [seemed] to be willing to give the defense counsel [Andy Anderson] anything and everything that [Anderson desired]," Zeller tried to get Bogoshian fired and replaced.[64]

Lieutenant Zeller complained in simple language to Rear Admiral

---

62 Appendix Ten: Zeller's memo to Bitoff of 23 November 1989, Thanksgiving Day.
63 Appendix Ten: Zeller's attorney work product memo to Bitoff of 23 November 1989.
64 Appendix Ten: Zeller's attorney work product memo to Bitoff of 23 November 1989.

Bitoff in a secret internal memo, written and submitted on Thanksgiving Day 1989, that Government Counsel Bogoshian (GC—a court-martial consultant) was not following orders and more egregiously lacked "the dedication of someone who desires to win."[65] Accuser Zeller was "sincerely convinced that [Prosecutor Bogoshian] did not have the desire to put the effort into this case which [was] required" and was as concerned with Bogoshian's demonstrated "naiveté" regarding witness manipulation. Zeller continued:

> [...] it seems as though [...] [Lieutenant Bogoshian lacked] not only experience, but also desire. One can be overcome by the other, but the absence of both leads to an untenable position.
>
> [...] it is recommended that corrective action be taken immediately to assign a special prosecutor to this case that will give it the attention it merits.

Obviously, Zeller opined, the case of XO *MARS* was as serious as it was complex. "I brought the charges and I convened the court-martial."[66]

Professor Everett makes note that one of the main features of the UCMJ "is that no commanding officer should try to influence courts-martial in any way, and none should censure or reprimand a court or its members for any judicial action taken."[67] Clearly, as the facts tell us, this UCMJ prohibition was inconsequential and not applicable for Bitoff and Zeller; further, the prosecution of the XO *MARS* warranted special handling and procedures, and not just a compliant special prosecutor. Intervention by Bitoff, Edwards, Zeller, the "command influence factor"[68]—even though such action is a criminal offense[69]—was

---

65 Appendix Ten: Zeller's attorney work product memo to Bitoff of 23 November 1989.
66 Appendix One: Bitoff letter of 30 April 1999.
67 Everett, p. 12.
68 Navy TJAG Grant's letter to US Senator Murray of 14 July 1994.
69 Under federal and military laws: Uniform Code of Military Justice under Articles Article 37(a).

unavoidable if Group ONE officers were to achieve what Navy TJAG Rick Grant later described as their "goal of a full and fair hearing"[70] for XO *MARS*.

Zeller, by way of his Thanksgiving missive, cemented and memorialized the attorney-client relationship that existed between Bitoff and Zeller, as it made clear that the communication up the chain of command was to be guarded as "ATTORNEY WORK PRODUCT."[71]

For reasons known only to Rear Admiral Bitoff and Lieutenant Zeller, and despite the best efforts of Lieutenant Matt Bogoshian (who Bitoff allowed to stay on as prosecutor), Zeller's investigation reports remained secret. Remarkably, the *USS MARS* fiscal year 1988 MWR report, collected and locked up by Zeller, was not presented as evidence! Even more amazingly, accusers Zeller and Bitoff did not testify against the accused; moreover, Bitoff's identity as accuser also went undisclosed! Throughout the proceeding, clandestine internal communications amongst Group ONE officers remained in the shadows.

The platoon of coerced[72] and prepared witnesses Bogoshian and Zeller marched into the hearing room was unable to make up for the total absence of physical evidence. As such, the Article 32 hearing officer Bitoff and Zeller had handpicked "to ensure that the complexity of the case (would) be appreciated"[73] found that "no reasonable grounds [existed] to believe the [XO *MARS*] committed the offenses [Bitoff and Zeller] alleged."[74]

Thus, Lieutenant Commander JJ Quigley found no evidence supporting a court-martial. Quigley attempted to persuade Rear Admiral Bitoff of the wisdom of that course of nonaction. In the absence of evidence, Quigley suggested there was nothing in Zeller's work, nor in his

---

70 Navy TJAG Grant's letter to US Senator Murray of 14 July 1994.
71 Appendix ten: Zeller's 1989 memo to Bitoff of 23 November 1989.
72 RDML Bitoff and Lt. Zeller's immunity grant to Lt. Brain Feely is discussed below.
73 Zeller's 1989 memo to Bitoff of 23 November 1989.
74 From Lieutenant Commander J. J. Quigley's Article 32 investigation report of 9 January 1990.

later pronouncement of guilt, to support a court-martial of any sort, and he therefore pointedly recommended that Bitoff not create (or convene) a court-martial.

Quigley noted, gently telling both Bitoff and Zeller, that there was no crime—not even a military crime—in honoring a fallen comrade by attending his funeral.

⚓

However, Quigley did recommend that Bitoff stand XO *MARS* before a "nonjudicial punishment" (NJP) disciplinary hearing (under UCMJ Art. 15, discussed earlier), which begs several questions. Among them: Why would Quigley think XO *MARS* would agree to a nonjudicial forum? Neither Bitoff nor the accused were serving aboard ship, so the vessel exception was not in play. Said another way, Bitoff was not in a position to force a NJP hearing; he lacked that authority. Another puzzlement was logical puzzlement: What did Quigley think Bitoff should examine as a military offense if there was no evidence to support Zeller's accusations, those already vetted by the Article 32 hearing?

Quigley's Article 15 recommendation informs the reader of the extraordinary pressure Quigley then faced, unable to deliver what Bitoff and Zeller had originally wanted. Further, applying Quigley's logic suggests the larger problem that an abusing commander faces against subordinates when using the NJP structure. That NJP was possible in a circumstance Quigley himself recognized as oxymoronic (since there was no incriminating evidence), and thus the entire situation gives us more reason to inquire more deeply into the world of nonjudicial punishment.

Predictively, Bitoff answered Quigley's recommendation by ignoring it.

At that point—and as far as XO *MARS*, this writer, was concerned—the questions regarding MWR disbursements had been fully resolved, and therefore I believed I would soon be released to carry out orders to report to the Naval War College in Newport, Rhode Island.

⚓

The court-martial for XO *MARS* began on February 3, 1990. Orders to the War College had already been quashed.

Bitoff and Zeller pressed ahead full speed to carry out their predetermined and vindictive court-martial. By so doing, Bitoff and Zeller left a paper trail as unavoidable as it was unmistakable, joining themselves at the hip to prosecution and defense attorneys in a scheme to rig a disciplinary hearing, one designed from the outset to punish and destroy an innocent man.

According to Zeller, "[as] an Article 32 UCMJ investigation had already been conducted, [RDML] Bitoff referred the case to a [special court-martial—SPCM], deciding against a [general court-martial—GCM] as [RDML Bitoff] did not desire to penalize [XO MARS] for exercising his right."[75] However, this incoherent explanation of Bitoff's decision does not make any sense! The more probable decision-making process in Bitoff's selection was first that the SPCM forum requires only three panel members, not five as for a GCM. The practical impact for Bitoff was that he would have to pick only three members from his personal Group ONE staff to sit in the court-martial panel. Second, special courts-martial draw less attention while still being fully capable of achieving the same "terrible consequences" that GCMs can garner.[76]

Military plantations are tight-knit and very small. By the time Rear Admiral Bitoff had decided upon a special court-martial for XO *MARS*, word had spread all over navy commands in the San Francisco Bay Area. The scuttlebutt was most certainly reverberating throughout Group ONE headquarters, as it also echoed off the bulkheads.[77] The rumor mill

---

75 Lt. Zeller's naval message to Capt. Glen Gonzalez of 16 April 1992.
76 RDML Bitoff's letter to Congressman Norm Dicks of 30 April 1999.
77 A bulkhead to a sailor is a wall to a civilian.

was electrified in headquarter cubicles occupied largely, according to Zeller, by officers like Steve Letchworth who had had numerous run-ins with XO *MARS* over one issue or another.

Unsurprisingly, it was from this tainted population pool that Bitoff selected his panel members, including Aviation Officer Letchworth—the same officer that Bitoff had disqualified for personal animus just a few months earlier as a potential investigating officer to the same case. By April 2, 1990, when the special court-martial met for the first time as a collected whole, Bitoff and Zeller (two accusers who had already declared a guilty verdict) had quietly put in place the prosecuting attorney, the defense attorney, and all three panel members. By receiving debriefs from JAG lieutenants that Zeller had dispatched to the hearing room, Zeller and Bitoff monitored the performance of this sycophantic claque daily. Zeller was also talking to the panel members about the hearing, thereby gathering information used to brief Rear Admiral Bitoff.[78] This Zeller was the same man who had declared the accused guilty at least twice before the disciplinary forum and who had been primarily responsible for the court-martial's creation.

⚓

Lieutenant Brian Feeley, who was on active duty serving aboard the *MARS* during the days surrounding Bill Nordeen's tragic assassination and had formerly acted as the *MARS* MWR account manager, left the navy before the *MARS* deployed in September 1988. Rear Admiral Bob Toney commanded Combat Logistics Group ONE at that time, not Bitoff.

When the *USS MARS* returned in March 1989, Brian Feeley was a civilian graduate student at Cornell University in Ithaca, New York. In chapter three of his book *Persons Who Can Be Tried By Court-Martial*,

---

78 Appendix Four: Zeller's "PERSONAL FOR" memorandum to Bitoff dated on or before 11 April 1990 as dated by allied papers.

Professor Everett points out that a guy like Brian Feeley, who had made a clean break from the military, was not subject to martial law. Since he was not "serving with, employed by, or accompanying the Armed Forces outside the continental United States" as a civilian employee, he was subsequently was no longer subject to Rear Admiral John Bitoff's command authority.

That legal nicety notwithstanding, Bitoff's legal advisor Zeller illegally threatened Feeley with court-martial if Feeley did not testify against XO *MARS* as a Bogoshian witness.[79]

Lieutenant Commander Quigley considered Mr. Feeley's immunized testimony during the Article 32 hearing before determining that charges against XO *MARS* were unsupportable.

Despite these obstacles, the Bitoff-Zeller team advanced charges in the court-martial.

Here is the linchpin: Bitoff and Zeller's chilling, impermissible, and illegal extension of the military establishment's discipline system into the civilian community clearly illustrates the acidic corrosion that military government has brought to pass upon our constitutional government.

⚓

To say that the 1990 special court-martial of XO *MARS* was rigged grossly would be a massive understatement of the actual documented events. "[Rear Admiral John Bitoff] brought the charges, and [Bitoff] convened the court-martial."[80]

As is routine for courts-martial, the XO's accusers did not testify in their capacity as accusers. Zeller, Romanski, and Bitoff did not appear in the hearing room at all. Again, for emphasis, Bitoff's identity as an accuser was a tightly held secret until, a whopping nine years later,

---

79 Bitoff signed Feeley's immunity grant on 14 November 1989. Also see Zeller's memo to Bitoff of 9 November 1989.
80 Appendix One: Bitoff's letter to Congressman Dicks of 30 April 1999, p. 5.

Bitoff first publicly disclosed his illegal court-martial participation to Congressman Norman D. Dicks (a US representative from Washington's 6th Congressional District serving as a Democrat from 1977 to 2013) in a letter dated April 30, 1999.[81] Bitoff again identified himself as the XO's accuser five weeks later, writing this time to Navy Secretary Richard Danzig with a copy also sent to the Judge Advocate General of the United States (Navy TJAG).[82]

For their part, both Edwards and Romanski remained cloaked and hidden as accusers for over twelve years until the release of internal Group ONE correspondence.[83]

⚓

Captain Mike Nordeen signed out a formal accounting of the *MARS* MWR account for fiscal year 1988 in late September 1988. Both the original and copies were mailed to the immediate superior in the chain of command, Rear Admiral Bob Toney, as required, just a week after the ship deployed. Requests for additional copies were promptly obliged. Copies were maintained aboard the *MARS* in compliance with normal record-keeping procedures.

Expenditures for the year were professionally detailed in an item-ized report prepared by supply officers working for Commander Arthur Rorex (CDR and a navy O-5), who reviewed the report before it went to Captain Nordeen for signature. Commander Rorex was supply officer aboard the *USS MARS*, trained and experienced in such affairs of finan-cial management.

Lieutenant Zeller obtained at least one copy of the *MARS* MWR annual report for FY-1988 while we embarked from September 18 to

---

81 Appendix One: Bitoff's letter to Congressman Dicks of 30 April 1999.
82 Bitoff's letter to Navy Secretary Richard E. Danzig of 4 June 1999.
83 Zeller's internal memo to Paul A. Romanski of 2 November 1989. Also Romanski's 3 No-vember 1989 "MEMORANDUM FOR THE RECORD, Chronology of Events."

26, 1989.[84] Later, Captain Edwards ordered that all *USS MARS* MWR records for fiscal years 1988–89 be turned over to Lieutenant Zeller so that all copies of the annual report came into Zeller's possession. Afterward, all copies were then locked away.[85]

In what was later realized to be subdued and secret discovery, Zeller released his formal investigation report of October 23, 1989, to Bogoshian and Anderson, government and defense attorneys respectively. There, on page one under paragraph five, entitled "Evidence Examined," is listed the Fiscal Year '88 MWR report for the *USS MARS*.[86]

⚓

I do not need to report on the hearings—neither the Article 32 hearing nor the court-martial.

A sequel to this book could be written discussing the fallacies and mendacious intentional misrepresentations that Zeller laid out in his pre-hearing investigation narratives, oral or written. Had Zeller's series of reports been made public in late 1989, his prevarications would have abruptly ceased; in fact, all stalking of and accusations upon this lieutenant commander would have become impossible.

To state that the two hearings—the Article 32 and the special court-martial—were rife with logical and legal problems seems redundant. To suggest that the main action of attack against XO *MARS* occurred in a public forum is simplistic. Burdening readers with detailed narratives of hearing room activities, trotting out witness lists, selectively quoting recorded testimony, parsing and analyzing the testimony, and poring over itemized lists of the documents that did find their way into the hearing

---

84 Zeller's voluntary sworn statement to NCIS Special Agent Dayne R. West of 21 February 1990.
85 Captain Edwards "PERSONAL FOR" naval message to Captain W.W. "Bear" Pickavance. "Bear" Pickavance relieved Captain Mike Nordeen of command on the *MARS* on Thursday, August 31, 1989. Kim Boyer's sworn statement.
86 Appendix Three: Zeller's investigation report of 23 October 1989, as endorsed by Captain Edwards on 26 October 1989.

rooms would be as tedious as it would be distracting. These aspects are, laconically, beside the point, as they are not useful.

Preceding and succeeding outside events sufficiently discredit everything that occurred within those two forums. For that keen reader who wishes to learn more, for either the bored or the curious, the Navy TJAG holds public records of both gatherings. But the kernel of truth about what actually took place is not preserved there. Nothing new was introduced at court-martial that had not already been explored during the requisite Article 32; only the list of players changed and grew.

Illegally stacking the deck, John Bitoff and Tim Zeller picked all their court-martial panel members from Bitoff's own personal Group ONE staff.

⚓

Professor Everett writes, "Review of a case will begin with the preparation of an appropriate record of trial[87] and its transmittal to the commanding officer who convened the court-martial."[88] A senior commanding officer is required to carefully consider consultant recommendations devolved from a court-martial disposition before acting, in the name of the commander in chief, to exonerate or punish the person accused.

Legal advisor, advocate, and investigator Tim Zeller pronounced XO *MARS* guilty three times months before the court-martial. Subsequently, Bitoff compelled the court-martial to take place; he also arranged the assembly to guarantee that Zeller's verdict would be formally repeated during the court-martial. Bitoff did not need a hearing record to declare his forgery authentic. Bitoff acted to punish XO *MARS* a week after the hearing.[89] It was another three weeks before the hearing record was

87 Professor's Everett's deception, not mine. Courts-martial are not federal trials.
88 Everett, p. 260.
89 Zeller's secret memorandum to Bitoff of 11 April 1990, "FINAL DISPOSITION OF MWR ACCOUNTABLE PARTIES."

authenticated and made ready for review.[90]

Speaking of forgery...

Army TJAG Samuel Ansell, as earlier noted, wrote an excoriating analysis of military discipline, a work published in 1920. General Ansell writes, "It helps the investigating officer to impose his authority upon the unfortunate suspected man and enmesh him in words and conduct having no origin in fairness and truth."[91] The tactic of pressuring an accused to convict himself extends beyond investigating officials to any government agent whose job it is to win—er, I mean "attain." For men such as Bitoff, Edwards, and Zeller, it is "axiomatic" they "want to get at the facts [no matter how] for the sake of discipline [...]. There is no better witness against a man than himself"[92] (my own emphasis added).

Beyond recognizing the "How?" component of getting at the facts, we must also consider the remaining elements in the classic reporting model answering these basic questions: Who? What? When? Where? Why? And yes, indeed: How?

Who: An accused soldier or sailor in a system not governed by law.

What: Forcing those accused to testify against themselves.

When: Whenever necessary. Each event is tied to a discrete ad hoc court-martial, before, during, or after.

Where: Wherever a court-martial is created (convened).

Why: To maintain good order and discipline.

How: Several ways exist to ensure that government agents are able to force a confession.

For example, here's one: Forge the accused's confession.

Almost immediately after the hearing ended, consistent with ongoing exertions to deflect attention away from his vindictive actions and those of his cohorts, John Bitoff commissioned the fake confession of XO

---

90 Notice of authentication of the hearing record of 5 May 1990 contained in the record.
91 Ansell, p. 13.
92 Everett quoting Napier, p. 12.

MARS. And, keeping all matters in-house, that assignment went to the same man Bitoff and Zeller had personally selected as defense counsel: Captain of Marines Kevin M. "Andy" Anderson!

Tim Zeller's wife was a navy nurse in July 1990. Speculation runs that Maureen Zeller was working at the Oak Knoll Naval Hospital in Oakland, California, when the XO's wife was admitted for a scheduled Cesarean delivery of XO's youngest daughter. The suspicion runs further that Maureen Zeller confirmed the birth to Tim Zeller the night of July 6.

It is known with certitude that Rear Admiral Bitoff, through Tim Zeller, delivered to Defense Counsel Andy Anderson the XO's punitive letter of reprimand the next day, July 7.[93] What is more, Anderson was told in writing that his client, the XO, had ten days to respond to the punitive letter should he desire to do so.

Defense Attorney Anderson, working for Bitoff and Zeller, authored the XO's confession inside the next ten days. Then, Anderson gave the criminal instrument to Zeller.

No board or review ever saw or studied the aftermath of Bitoff's, Edwards's, and Zeller's dastardly handiwork.

With this vile deed accomplished, John Bitoff quietly retired in October of 1991.

Suddenly, in a flash, any chance I might have had to attend the War College in Newport, Rhode Island, and to thereafter advance in rank was unceremoniously rescinded, dashed, truncated.

I do thank the reader for staying with me and reading the entire story of this biased military trial knowing full well that is it a grimy and dirty tale, one full of tortured complexity and obtuse machinations.

---

93 Court-martial consultants to RDML Bitoff recommended, and Bitoff ordered XO *MARS* punished through issuance of a career-ending letter of reprimand.

7.
HOW WE GOT HERE!

"The way to have good soldiers
is to treat them rightly [...].
A private soldier has as much right
to justice as a major general."

—Abraham Lincoln

Written for the benefit of John Bitoff...

Since adoption of the military code in 1776, America is running

on three separate governments, all without God!

⚓

AMERICA, THIS IS WHERE WILLIAM WINTHROP (1831–1899) got his idea.

Prior to the ascent of Henry II (1154–1189) to the throne of England (he was also known as Henry Curtmantle or Henry Plantagenet), manor barons unevenly "dispensed justice whose quality and character varied with the custom and temper of the neighborhood."[94]

For his part, King Henry Plantagenet carefully and gradually introduced a system of royal courts tasked with administering a law common to "all England and all men." Here we see the birthplace of the English common law.[95]

---

94 Churchill, Winston Spencer. (1956) *A History of the English Speaking Peoples: The Birth of Britain (Vol. I)*, New York, Dodd, Mead & Company, p. 216.
95 Churchill, Winston Spencer. (1956). *A History of the English Speaking Peoples, The Birth of Britain (Vol. I)*, p. 216.

Many common law practices were unwritten and unevenly conducted. Years would pass for the diversity of the royal courts throughout England, along with local community and clerical influences, to disintegrate so that thereafter, the rules would become common and universal throughout the country.[96]

About the year 1250 a Judge of Assize named Henry of Bracton produced a book of near nine hundred pages entitled *A Tract on the Laws and Customs of England.* Nothing like it was achieved for several hundred years, but Bracton's methodology set an example, since followed throughout the English-speaking world, not so much stating the Common Law as of explaining and commenting upon it; and thus encouraging and helping later lawyers and judges to develop and expand it.[97]

After America's Civil War and his retirement, Acting Judge Advocate William Winthrop, as a life's purpose, took it upon himself to set down a comprehensive treatise "on the science" of what Winthrop called "military law,"[98] a term more properly described in this book as "military rule."

Winthrop's dedicated goal was to maintain the Articles of War and the Articles for the Government of the Navy as entities operating separately from American's common law government under the Constitution during peacetime. This crucial sleight of hand cloaked the quality of the military rule, making it appear to be constitutionally acceptable and exemplifying the quality of government under a representative republic.

Understandably, by 1861, many military discipline practices were unwritten and unevenly conducted. Ashamed and disgusted by unprofessional and cruel military discipline practitioners, unwarranted

---

96 Churchill, Winston Spencer. (1956). *A History of the English Speaking Peoples, The Birth of Britain (Vol. I)*, p. 216–224.

97 Churchill, Winston Spencer. (1956). *A History of the English Speaking Peoples, The Birth of Britain (Vol. I)*, p. 224.

98 Winthrop, William. (1886). *Military Law and Precedents, Vol. I.* Washington, D.C.: War Department No. 1001, Office of the Adjutant General, p. 5.

courts-martial, inconsistencies in courts-martial outcomes, and military prosecutions of civilians, Winthrop adopted Henry Bracton's methodology, creating single-handedly a treatise on the full scope and operation of the military discipline system. Winthrop falsely claims that the military law derives from common law, yet it has also a *lex non scripta,* or unwritten common law, of its own.[99]

The key fact is that Winthrop wanted to alter the separateness of military rule from civil law; thus, he tried to make it look as though military personnel enjoyed the same rights as their civilian counterparts.

Winthrop writes in the 1886 first edition preface of his weighty 1,111-page doorstop tome:

> In view of the absence and want of a comprehensive treatise on the science of Military Law, it has been for some years the purpose of the author—a member of the bar in the practice of his profession when, in April, 1861, he entered the military service—to attempt to supply such a want with a work, which, by reason of its extended plan and full presentation of principles and precedents, should constitute, no merely a text book for the army, but a *law book* adapted to the use of lawyers and judges.[100]

In the course of writing his work, Winthrop shifted and assimilated unwritten military traditions, coalesced services, rules, regulations, disciplinary standards, portions of the martial code, and arbitrary exercises of military authority into one tome, one designating "the custom of war." And thus, one single man rebranded and repurposed all of these diverse elements into "the military common law!"[101]

99 Winthrop, William. (1886). *Military Law and Precedents, Vol I.* Washington, D.C.: War Department No. 1001, Office of the Adjutant General, p. 41.

100 Winthrop, William. (1886). *Military Law and Precedents, Vol I.* Washington, D.C.: War Department No. 1001, Office of the Adjutant General, p. 5.

101 Winthrop, William. (1886). *Military Law and Precedents, Vol I.* Washington, D.C.: War Department No. 1001, Office of the Adjutant General, p. 41. Also see: Kastenberg, Joshua E. (2009). *The Blackstone of Military Law: William Winthrop.* Lantham, Maryland: Scarecrow Press Inc., p. 231.

Moreover, Winthrop, using clever wordplay, attempts to pass off the military code as establishing its own cultural "common law" and sinisterly makes out his military "common law" as a descendent of, and one on equal footing with, the English and American common law.

Herein revealed is Winthrop's Curse, what I term the "durable myth"!

Whereas Winthrop elsewhere concedes the military custom of war and America's common law serve two distinct and independent government sovereignties of the United States and separate states, serving two separate and distinct functions—one governing in time of war, the other governing in time of peace—Winthrop's Curse strives to eradicate those distinctions and graft one sovereign onto the other. His ultimate aim is to make it appear to the reader that the two are as one, working in some sort of faux, disjointed harmony with each other.[102]

Therefore, we may consider military traditions and customs as akin to a building structure of a distinct and unique architecture, one unlike any other.

Then consider America's republican form of government as an altogether different edifice, unique and unlike any other government structure.

Winthrop's Curse advances and nourishes the durable myth that both government buildings were constructed using the same set of blueprints...

But that is impossible!

Intentionally, Winthrop seeks to hide the reality that the common law—that is, our constitutional law—is the supreme sovereign of judicature in America. He wants us to believe that the constitutional civil laws and forms of proceeding regulate the practice of all other courts of justice in the United States, including over the military sovereignty. However, this is an obvious falsehood!

Please consider that rules are regulations for the military establishment; indeed, crucially, they are not laws, and they are not legislated

---

102 Winthrop, Vol II, p. 691.

nor made into statute by elected representatives. It ought to be again stated: The Uniform Code of Military Justice (UCMJ) and the Manual for Courts-Martial (MCM) are a set of agreed-upon rules, not laws, produced by independent, unelected legally trained military sovereigns tasked with creating, amending, and enforcing the military code. They are mandated![103]

Further, the UCMJ and the MCM are enacted and mandated by presidential executive order.

The result has been to give senior military tyrants an entirely free hand in the operation of their sovereign, arbitrary military authority, and therefore, if so inclined, they may ravage subordinates as animals unconstrained. These military leaders are unconcerned and unfettered by normal constitutional prohibitions. Their actions are all functions of command.

Thus, America is always at war within itself to appease the appetite of the insatiable wartime sovereign.

Winthrop writes: "[A court-martial] is not a [...] 'court of record.' [Courts-martial have] no power to punish for contempt, no power to issue a writ or judicial mandate, and its judgment is simply a recommendation, not operative till approved by [the] commander."[104] As observed by one British military discipline disciple, "It must never be lost sight of that the only legitimate object of military tribunals [courts-martial] is to aid the Crown [our commander in chief] to maintain the discipline and government of the army [and navy]."[105]

I recognize, and at the same time roundly condemn, Winthrop's tremendous success in achieving his goal: WINTHROP'S CURSE!

To fully appreciate what Winthrop accomplished in *Military Law*

103 Joint Service Committee on Military Justice: https://jsc.defense.gov. Also see: The Uniform Code of Military Justice and Manual for Courts-Martial: Retrieved from: https://mdwhome.mdw.army.mil/docs/media-documents/ucmj.pdf.
104 Winthrop, p. 50.
105 Winthrop, p. 50, n. 24.

and Precedents, let us step back to study the earlier foundational work of Judge of Assize Henry of Bracton (1210–1258), cleric and jurist.

"In all this time however only one man attempted to a general and comprehensive statement of the English Common Law. About the year 1250 [...] Henry of Bracton produced a book of nearly 900 pages entitled *A Tract on the Laws and Customs of England*. Nothing like it was achieved for several hundred years, but Bracton's method set the example, since followed throughout the English-speaking world, not so much as stating the Common Law as of explaining and commenting on it, and thus encouraging and helping later [civilian] lawyers and judges to develop and expand it."[106]

Army JAG Winthrop was one of those who followed Bracton's example.

Winthrop's view of the military discipline system was that it was unprofessional, in the Swaim court-martial, and other personal experiences (pp. 229, 231).

With a navy captain and JAG combined in the person of Commander Edward M. Byrne, an excellent history of the evolution of the military discipline exists entitled *Military Law* (Naval Institute Press, Annapolis, 1976).

Regrettably, to the everlasting detriment of every person who has ever served in the United States military, Winthrop and Byrne have constructed and greatly expanded a Potemkin village, one intended to fool outsiders that military discipline has a connection to justice. The original term "Potemkin village" derives from a story dating back to eighteenth-century Russia, suggesting that an artificial place can be built to disguise or conceal the true, and often less desirable, identity of the original. This tactical and intentional gambit so far has been supremely successful in pulling the wool over the eyes of the uninformed or disinterested; non-students of martial discipline think they see a village where there exists

---

106 Churchill, Vol I, p. 224.

only the cheap imitations, propped-up storefronts.

Military writers such as Winthrop and Byrne expound with great purpose and vigor that the Articles of War and the Uniform Code of Military Justice are found in our Constitution. These people, serving the best interests of the government (and their own personal interests) in their writings, create a public acceptance of the military discipline system as one that is compatible with the Constitution and as reliable as the federal judiciary. However, as we have seen, their interpretations are patently and intentionally false.

WILLIAM WINTHROP'S CURSE: WINTHROP AS PRINCE POTEMKIN

Methods devised in the practice of law in the formative years of Henry II's reign relied upon "unwritten custom of the land as declared by the inhabitants and interpreted, developed and applied by the judges." In the nascent days of the burgeoning English common law, littered with centuries-old and uneven rulings, those few reduced to writing were applied erratically, case by case, and in an unprofessional and undisciplined fashion.[107]

THE ART OF ASSIMILATION—RESISTANCE IS FUTILE!

Assimilation, in the context of the present argument, means first to make military governance look similar to constitutional government and second to use simple trickery to absorb that foreign and culturally distinct martial government into the prevailing constitutional culture.

Colonel Winthrop recognizes that the function of a courts-martial is an "exceptional forum,"[108] very different from civilian courts, but he goes on to suggest close analogies to its *personnel* where there are really none. "Thus [according to Winthrop] courts-martial have frequently

---

107 Churchill, Winston, Spencer. *History of the English Speaking Peoples.* Volume I, pp. 221–224.
108 Winthrop, p. 54.

been compared, as to some of its powers and proceedings, to a *judge*, and as to others to a *jury*."[109] Telling the public there exists similarity in some respects between courts-martial and civilian courts—otherwise dissimilar in every respect—are intentional acts. Such a statement is an act of deception and defiance rather than an act of assimilation.

DISRUPTING DURABLE MYTHS

The basic principles of the United States Constitution are the separation of powers within a representative republic and federalism. However, these principles find no home in the military community, something which may come as a shock to many. Some observers are vaguely aware that the Constitution does not protect soldiers and sailors; however, they are hard-pressed, when put upon, to articulate precisely what protections members of the armed forces forfeit. Pointedly, many Americans cannot answer why military men and women should have to give up any constitutional protections in the first place.

For instance, military affairs correspondents covering courts-martial interview persons, described in subsequent news reports as "prosecutors," who argued against "defense lawyers" during a "trial" wherein a "fifteen-member jury"[110] handed down a "verdict" of guilt.

Military governors are greatly pleased and view reports like these as significant measures of success in a centuries-old campaign of creeping obfuscation. The real truth has become clouded. Many (or perhaps most?) civilians are actually convinced that there is a judiciary branch of the national defense establishment. Defense Department convening authorities, under orders of the current commander-in-chief, can walk any dog into a courtroom full of cats, and no one sees anything amiss.

Appointed unelected admirals and generals, through an agency operation between the commander-in-chief and his anointed subcommanders,

---

109 Winthrop, pp. 54—55. (Emphasis in the original.)
110 Thompson, Estes. 2005, April 22. Seattle Post-Intelligencer, p. A3.

enforce edicts intended to punish soldiers and sailors. This is an agency of military command governed and controlled by the commander-in-chief, as authorized by the Congress and sanctioned by Supreme Court justices.

Thus, we see a clandestine act of supremacy, and the US Supreme Court quietly sanctions the status quo: *The Federal Courts and Military Justice: To Intervene or Not?*[111]

⚓

John Adams gave his approval to this cunning scheme in 1776. It was the first time in history that two foreign sovereigns came into close combat, together bracing the same orders of battle.

> There was an extant, I observe, one system of Articles of War which had carried two empires to the head of mankind. The Roman and the British; for the Articles of War are only a literal translation of the Roman. It would be a vanity for us to seek our own invention or the records of warlike nations of [*sic*] a more complete system of military discipline. I was, therefore, for reporting the British Articles of War *totidem verbis [...]*.[112]

In so many words, in these exact words.

Unfortunately, in 1800, a motion to reverse the strictures of the Articles of War of the government was struck down by a large majority in Congress.[113]

---

111 Generous, William T. 1973. *The Federal Courts and Military Justice: To Intervene or Not?* Kennikak Press, p. 165.

112 Valle, James E. 1980. *Rocks and Shoals: Naval Discipline in the Age of Fighting Sail.* Naval Institute Press, p. 41. (This was the first time that two competing armies fought using the same Articles of War.)

113 Sullivan, Dwight H. December 1998. *Playing the Numbers: Court-Martial Panel Size and the Military Discipline*, p. 11.

8.
NO LAW AT ALL

⚓

FOR ANY VETERAN, WHETHER ON active duty or having completed prior service, no jurisdictional mandate should cause more consternation than this one:

> This provision, which, derive original from a corresponding British Article but a single material change, presently to be notice [English to American], since its first appearance in our code of 1776, proceeds upon certain general principles well defined on our law. Of these, the fundamental principles of the distinctness and independence of the two sovereigns of the United States and of the separate States [...]. But not withstanding of the military power within its peculiar field, the future principle is uniformly asserted of the subordinate, in time of peace and on the common ground, of the militarily authority to the civil, and of the consequent amenability of military persons, in their civil capacity, to the jurisdiction, for breaches of the criminal law of the land.

THE THREE SOVEREIGNS: ROME, BRITAIN, AND STEALTH AMERICAN!!

In plain language, what this means is that the constitutional law belongs to military laws.[114]

However, the truth is this: The history and character of the ordinary

---

114 Winthrop, William. 1896 (Reprinted 2000). *Military Law and Precedents*. Beard Books, p. 691.

laws and punishments, the constitutional law under our Constitution—
which presumably protects civilians at all times, in peace and in war—and
the source of the sovereign military government's jurisdiction is derived
from the sovereign constitutional law—our civil government.

⚓

Now we must go back and read those sentences again!

9.
WHY ARE WE STILL HERE?

⚓

"The discipline which makes the soldiers of a free country reliable in battle is not to be gained by harsh or tyrannical treatment. On the contrary, such treatment is far more likely to destroy than to make an army. It is possible to impart instruction and give commands in such a manner and such a tone of voice as to inspire in the soldier no feeling but an intense desire to obey, while the opposite manner and tone of voice cannot fail to excite strong resentment and a desire to disobey. The one mode or the other of dealing with subordinate's springs from a corresponding spirit in the breast of the commander. He who feels the respect which is due to others, especially his subordinates, cannot fail to inspire hatred against himself."

—Major General John M. Schofield (1831–1906)
His Address to the West Point Class of 1879

⚓

10.
MWR EXPENDITURE

⚓

MWR FUNDS ARE GENERATED FROM profits received from the sales of commercial goods aboard military reservations, be they base exchanges ashore, soda machines around squadron spaces at military air bases, or ship stores selling unit emblematics like baseball caps, belt buckles, or patches. Stores onboard ships also sell everyday consumables that are staples in everyone's life—toiletries, health and comfort products, food and "geedunk" (candy, ice-cream)—just like a 7-Eleven in your neighborhood. MWR monies are private. The dollars that build and contribute to a MWR fund are in no way connected to tax monies that our government collects from citizens and then appropriates to pay for our military operations worldwide.

MWR monies are "non-appropriated" funds. The money belongs to the personnel within any given military unit whose personnel are the only contributors paying into that fund.

The *MARS* MWR fund, for instance, came from profits from items our crewmen purchased on the ship—that is, *USS MARS* crewmen only.

MWR funds are to be used exclusively for the morale, welfare, and recreation of the unit members who pay into the account. Committees and councils are established to recommend, decide, administer, account for, and oversee MWR fund spending regarding the distribution of these *non-appropriated* dollars.

Only one charge forms Zeller's singular charge sheet deal with matters related to the *USS MARS* (AFS-1) minivan pick-up. None came forward to testify. Zero.

The charge was then dropped.

The charge I was "convicted" of came from a ghost who astoundingly testified before a rigged jury.

11.
JOHN BITOFF TAKES COMMAND

⚓

JOHN W. BITOFF WAS COMMANDER of the Combat Logistics Group ONE in August 1989. Bitoff, a one-star rear admiral, took command of CLG-1, relieving Robert Toney, in January 1989. The *USS MARS* was deployed overseas when Bitoff took the CLG-1 helm. Navy Captain Michael B. Edwards was one of Toney's, and then Bitoff's, chiefs-of-staff.

"In 1989, [Timothy W. Zeller] was Staff Judge Advocate for Commander, Combat Logistics Group 1. The Staff Judge Advocate is responsible for supervision of all legal matters within the Group and may be equated to the equivalent of a civilian in-house counsel and assistant district attorney combined."

As Mike Nordeen was in the last days of his command of the *USS MARS* (AFS-1), Bitoff and company wrongly alerted higher command, headquartered in San Diego, that "wrongdoing had been discovered" regarding the *USS MARS* MWR account. According to Zeller, "In the fall of 1989, [approximately] ten days prior to the commencement of [PACIFIC EXERCISE 1989], [Combat Logistics Group ONE] was tasked by Commander Naval Surface Force, Pacific Fleet to conduct an Integrity and Efficiency [I&E] investigation into irregularities of the Morale Welfare and Recreation account of *USS MARS* [AFS-1]."

Meanwhile, Bitoff and Zeller waited in ambush.

Captain Mike Nordeen handed over the keys to the *USS MARS* to William "Bear" Pickavance in a change of command ceremony on Thursday, August 31, 1989. Newly installed Commanding Officer

Pickavance immediately declared the next day, Friday, September 1, a ship's holiday—liberty for all hands save the duty section, which happened to kick off the Labor Day weekend.

It was that same Friday, the day after the change of command, that Rear Admiral Bitoff decided to commence the MWR audit of the *USS MARS* MWR account. However, the ship was deserted, manned only by a skeleton crew. Christopherson's inspection was hampered inasmuch as personnel were on liberty, thereby rendering a number of MWR records unavailable.

Bitoff's scheduling of the audit was highly abnormal and irregular. First, it was out of periodicity, several months late, and second, it was carried out on the *day after* a shipboard change of command, which was also the start of the Labor Day break.

Ruth A. Christopherson, headquartered in San Diego, was a civilian employee working for Vice Admiral Bennett as a subject matter expert in the operation of the MWR organizations throughout the surface fleet. Ruth had started on the bottom rung of the MWR department at fleet headquarters, eventually, by 1989, working her way to the top as the fleet's MWR auditor.

Rear Admiral Bitoff and his staff used Ruth Christopherson's MWR audit in a dedicated design to target Lieutenant Commander Fitzpatrick as a criminal suspect with respect to MWR expenditures.

Christopherson was alerted by a Bitoff staffer regarding the invented problems with the *MARS* MWR account, and that staffer directed Christopherson to focus intently on funeral expenditures from July 1988.[115] Setting me up for the ambush, Lieutenant William Bramer, assistant supply officer on Bitoff's detail, pre-briefed the MWR auditor in staff offices ashore before Christopherson even set foot aboard the *USS MARS.*

---

115 Lieutenant Bill Bramer's sworn affidavit.

⚓

By the way, the *USS MARS* (AFS-1) was named for the borough of Mars, Pennsylvania, which is named for the planet Mars.

In my over two-year tour as XO, the *MARS* MWR audit was the only inspection the ship underwent where I was not physically present aboard ship.

Believe it or not, I never met Ruth Christopherson.

Students from the Mars Middle School were pen pals with *USS MARS* crewmembers. When students were told of the upcoming August 1989 change of command, one of the middle school teachers, Peg Harding, asked for permission to attend the ceremony with one of her students and visit the ship. Captain Nordeen said, "Sure!"

A fund was set up in the city of Mars to cover trip expenses. Some of the *USS MARS* crew chipped in.

My family hosted their visit. I picked Mrs. Harding and her eighth grade student Jennifer MacDonald up at the airport the day before the change of command on Wednesday and drove them directly to the ship. Captain Nordeen had given the two women permission for an overnight stay aboard, which thrilled them to no end. Female *MARS* crewmembers chaperoned them during their time aboard.

On Thursday, after the CO handoff, Peg and Jennifer stayed at my home for the remainder of their visit.

On Friday, my family and I took Peg and Jennifer to visit the *USS ENTERPRISE* (CV-65) and then on a day tour of San Francisco, traversing the historic and scenic forty-nine-mile tour route.[116] While we were taking in the sights of San Francisco, Ruth Christopherson was conducting her MWR audit aboard the practically deserted *USS MARS*.

---

116 12 September 1989. Cranberry Eagle article.

⚓

In keeping with the preplanned result, Ms. Christopherson failed the MWR account managers in her pro forma report by pointing chiefly at funeral trip expenses as her chief complaint, thus parroting Lieutenant Bramer.

"The alleged irregularities involved an expenditure of approximately $10,000.00 to send certain members of the USS MARS crew to the funeral of the USS MARS CO's brother (Navy Captain William Edward Nordeen), a Naval Attaché who had been murdered by the terrorist group known as 'November 17.'"

Christopherson also called out misuse of funds for the Hawaii trip, although "it was known at that time that the executive officer Lieutenant Commander Fitzpatrick was intimately involved with the administration of the MWR account."[117] Putting aside questions about Ms. Christopherson's independence and objectivity, the experienced MWR examiner, with good reason, did not make a single mention—by name or position—of the executive officers of the MARS in her twelve-page audit findings. Shipboard executive officers, as was normal ship routine and organization, did not hold responsible positions in the management or oversight of MWR accounts.[118]

Christopherson then returned to San Diego and reported her MWR audit results to her boss, Vice Admiral Bennett. Bennett then turned around in his swivel chair and ordered Rear Admiral Bitoff to conduct an administrative Integrity and Efficiency report.

As a function of his command, Bitoff appointed his own staff attorney, Tim Zeller, as preliminary investigating officer in what was at this stage an administrative procedural step.

Bitoff and Edwards first revealed their criminal misconduct at this

---

117 Zeller's 8 July 1998 letter to the Oklahoma Bar Association (OBA).
118 Commander Fitzpatrick's FITREP does not identify any MWR duties.

important juncture: The detailing of the Integrity and Efficiency investigation officer was decided by Rear Admiral Bitoff and Bitoff's chief of staff, Captain Mike Edwards. This was a code meaning Bitoff was making believe he thought I had actually stolen the money used for the funeral trip and was setting me up for an ambush. In their arrangement, Bitoff and Zeller enjoyed an attorney-client relationship, an agency relationship that exists still today. When one man speaks, both men speak. When one man acts, both men act.

Zeller recounts Bitoff's decision this way: "Military case law has recognized the role as being both prosecutorial and as an objective reviewer at different times in the same trial. In the fall of 1989, I was tasked with conducting an investigation into the MWR expenditures onboard *USS MARS*, said tasking being a result of a directive from Vice Admiral Bennett. My client in this matter, as both an investigator and Legal Officer, was the Department of the Navy as personified by Rear Admiral John Bitoff, then Commander, Combat Logistics Group 1."

Originally, the plan was to send the Combat Logistics Group ONE staff officer assigned to Bitoff as the assistant training officer, Lieutenant Commander Steve Letchworth, and Lieutenant Commander Zeller, then a lieutenant. However, just prior to the detailing, it was decided by Bitoff and Edwards that Lieutenant Zeller would conduct the investigation alone.

As memory serves, this decision was made for two reasons:

First, Lieutenant Zeller was one of the few officers on the Combat Logistics Group ONE staff who had not had a confrontation with Lieutenant Commander Fitzpatrick, the executive officer. Second, with Combat Logistics Group ONE immersed in Pacific Exercise 1989, other options were limited.

These factors were considered sufficient to override normal protocol that a junior not investigate a senior.

Thus, Lieutenant Timothy W. Zeller (0-3), Bitoff's staff JAG, commenced his assigned duty on 18 September 1989.

⚓

Lieutenant Zeller departed onboard the *USS MARS* to conduct the Integrity and Efficiency investigation while the ship was en route to Dutch Harbor, Alaska.

12.
YOU CAN HEAR THESE GUYS TALKING: SUMMARY JUDGMENT!

⚓

"'Herald, read the accusation!' said the King.

On this the White Rabbit blew three blasts on the trumpet,

And then unrolled the parchment-scroll, and read as follows:

'The Queen of Hearts, she made some tarts,

All on a summer day:

The Knave of Hearts, he stole those tarts

And took them quite away!'

'Consider your verdict,' the King said to the jury.

'Not yet, not yet!' the Rabbit hastily interrupted. 'There's a great deal to come before that!'"

~

"'Consider your verdict,' the King said to the jury.

'Not yet, not yet!' the Rabbit hastily interrupted. 'There's a great deal to come before that!'

'No, no!' said the Queen. 'Sentence first—verdict afterwards.'

'Stuff and nonsense!' said Alice loudly. 'The idea of having the sentence first!'

'Hold your tongue!' said the Queen, turning purple."

—Lewis Carroll (1832–1898), 1865, *Alice's Adventures in Wonderland* and *Through the Looking Glass*

⚓

IN A PRELIMINARY STATUS REPORT on the inquiry, Bitoff reported to Vice Admiral Bennett that even though it was suspected that an executive officer afloat, Lieutenant Commander Fitzpatrick, had stolen over $10,000 from his *MARS* shipmates out of the command MWR account, Bitoff was going to press on unilaterally with the inquiry—an action done without first notifying the Naval Investigative Service (NIS) as required.

So, rather than contacting the NIS, which Bitoff says he personally dismissed out of hand because the *MARS* was going to deploy, Zeller recounts that "the detailing of the Integrity and Efficiency investigation officer was decided unilaterally by Rear Admiral Bitoff and Bitoff's Chief of Staff, Captain Edwards."

Military case law has recognized this role as being both prosecutorial and an objective reviewer at different times in the same trial.

⚓

It is assumed that Captain Edwards hatched the MWR plot against the author in the days after Edwards's temporary command of the *USS MARS*, carried out upon his return to CLG-1 staff duties in July 1988.

Zeller reports: "Upon completion of the investigation Lieutenant Zeller departed the ship in Dutch Harbor and flew to the Naval Shipyard Puget Sound where the advance base of Combat Logistics Group ONE was located for PACIFIC EXERCISE 1989."

An initial debriefing was delivered to Rear Admiral Bitoff, and if the recollection is correct, Captain Edwards was present.

The assertion by Lieutenant Commander Fitzpatrick concerning Captain Edwards's alleged acquiescence or approval of the plan for the funeral party was, to the best of Zeller's recollection, not discussed.

If memory serves, this initial debrief was cursory for two reasons.

First, though the material and interviews had been completed, they had not yet been transformed into a written report with conclusions and recommendations. Second, the more immediate (Combat Logistics Group ONE) command concern was an unrelated officer misconduct case, which had been handled by the investigating officer for another Combat Logistics Group ONE ship during the transit to Dutch Harbor.

Civilian and uniformed federal authorities have continued to hold me as the target of their federal felony criminal enterprise each and every day for what, as of this writing, exceeds thirty-one years.

I cannot speak with precision about when the criminal plot against me began. Only men like John Bitoff, Mike Edwards, Paul Romanski, and Tim Zeller know the complete truth, yet all remain hidden and silent. Perhaps one day, more incriminating documentation shall be found and brought to light.

Let us step back a little and consider for a moment some penetrating and pertinent words: Cavorting. Unconscionable. Conspiracy.

con·spir·a·cy, *noun, plural* **con·spir·a·cies.**

1. the act of conspiring
2. an evil, unlawful, treacherous, or surreptitious plan formulated in secret by two or more persons; plot
3. a combination of persons for a secret, unlawful, or evil purpose [k*uh***n-spir-***uh***-see]**

Clearly, Bitoff's criminal conspiracy was underway well before September 1989. It would be good to know precisely when the machinery of this outlaw industry began, yet it is likely not possible to know with certainty the exact date of commencement.

Meantime, there exist clear guideposts and mile markers, tracking backwards, that reveal Bitoff's plot. Such a study may move any serious investigator closer to the truth. In any case, progress forward can certainly be measured.

Let us consider the many players (co-conspirators) for today's performance:

- **Rear Admiral John W. Bitoff:** Commander of a group of a fleet of logistics ships operating in the waters between the International Date Line to America's western coast (Staff Code 00—"zero zero").

- **Navy Captain Mike Edwards:** Bitoff's second-in-command. Bitoff's executive officer, Bitoff's chief of staff (Staff Code 01—"zero one").

- **Navy Captain Paul Romanski:** An assistant chief of staff to Bitoff (Staff Code 02—"zero two").

- **Navy Lieutenant Timothy W. Zeller:** Bitoff's staff judge advocate general (Staff Code 14—"one four").

Pacific Fleet Exercise 1989 (PACEX '89) was the largest seagoing peacetime exercise in Pacific Fleet history up to that point in time.

Zeller and Romanski worked together assiduously on the illicit plot during that period of time, when most of the energies of Bitoff and Edwards were forced, focused, and directed toward an enormous and ongoing operational fleet exercise.

It is obvious that Paul Romanski had both the time and the industry to apply and craft a staff timeline chronology to memorialize the part that each man would play in the plot. Romanski filed the history as a **"MEMORANDUM FOR THE RECORD" on 3 November 1989.**

Romanski's memo reveals much. In and of itself a criminal instrument, Romanski's memo was kept secret from me from 3 November 1989 until 28 July 2001—a long period of time amounting to eleven years and nine months.

Romanski began the draft of his staff's illicit history sometime on or before 2 November 1989, a Tuesday.

Romanski worked with Tim Zeller along the way. This coordinated action comprised the aforementioned definition of conspiracy:

an evil, unlawful, treacherous, or surreptitious plan formulated in secret by two or more persons; a plot, a combination of persons acting for a secret, unlawful, or evil purpose.

Some of what the two conspirators left out is provided here (extracted from documents available upon request):

- **Monday, 18 September 1989:** Zeller embarked aboard the *USS MARS* to conduct his preliminary investigation.

- **Monday, 25 September 1989:** Zeller debarked the *USS MARS* on Monday

- **Exact Date Unknown:** Zeller becomes one of my accusers. Zeller orally briefs Bitoff and Edwards where they are forward deployed in Bremerton, Washington, aboard the Puget Sound Naval Shipyard. Zeller relates to Edwards and Bitoff that the case against me is serious. To wit: "Fitzpatrick is a guilty man!"

- **Wednesday, 4 October 1989:** Zeller›s status report is sent to higher command regarding the Naval Investigative Service. Truth is, no attempts were made nor was actual contact enjoyed between Zeller or Romanski to notify the Naval Investigative Service from the time of Zeller's return to Oakland (2 October) until the date, two days later, of Zeller's status report. As indicated below, Bitoff was back in San Francisco as of 4 October. (Please note: The organization's name Naval Investigative Service [NIS] did not change to Naval Criminal Investigative Service [NCIS] until 1992.)

Romanski did record this:

- **Monday, 2 October 1989:** Zeller returns to staff headquarters in Oakland, CA. Zeller orally briefs Paul Romanski. Romanski orders Zeller to prepare a written report. Once again, Fitzpatrick is a guilty man!

- **Wednesday, 4 October 1989:** Rear Admiral Bitoff returns to the San Francisco Bay Area. Bitoff is in town when Zeller, only back just two days himself, sends out his first preliminary investigation status report to higher command, stating that the NIS had been contacted but he had been too busy to answer the call!

- **Thursday, 5 October 1989:** Zeller has the first draft of his investigation report ready to show Romanski in three days' time. This is an extensive, carefully crafted report. Romanski and Zeller talk. During Romanski's verbal exchange with Zeller, Romanski reviews Zeller's written report. Romanski changes (or edits) Zeller's report sufficiently enough to join Zeller as accuser #2. By this time, both Zeller and Romanski are aware—both of them together—that they are accusers!

- **Tuesday, 10 October 1989:** Romanski briefs Edwards on the investigation report that by now both Zeller and Romanski have authored and edited (with Romanski and Zeller acting as accusers). In separate writings, and in Zeller's own words, Mike Edwards interacts with Zeller and edits Zeller's investigation enough that he may join Zeller and Romanski as another accuser; thus, Edwards becomes accuser #3. (Please consider: Edwards could have become an accuser when Zeller briefed him in Bremerton, Washington, in late September. At the time of writing, I simply do not have enough clear evidence to know for certain.)

- **Thursday, 12 October 1989:** Zeller briefs Rear Admiral Bitoff, Captain Edwards, and Captain Romanski in Oakland, California. This is the conduct of the cloaked court-martial. Zeller repeats his pronouncement that "Fitzpatrick is a guilty man!" Bitoff, Edwards, and Romanski extend themselves in their broad concurrence. Arrangements are immediately commenced for the SHOW—the more public court-martial. Amazingly, Zeller's preliminary investigation report at this moment exists only as a DRAFT!

Bitoff ordered Zeller to advance formal charges against me after the 12 October 1989 secret court-martial. "I brought the charges, and I convened the court-martial," Bitoff admits in writing in a letter sent to Congressman Norm Dicks and Navy Secretary Danzig dated 30 April 1999.

The kind reader well knows that accusers are supposed to be prohibited from "convening" (creating, assembling, gathering, conducting) in secret and in advance of their own courts-martial.

However, in a world where the rules, regulations, and laws are meaningless, in a world without real consequences, you find menaces like John Bitoff and his band of merry men engendering and exploiting a system of tyranny to the hilt.

So, **it was only after Bitoff had given Zeller the green light** to proceed and after he had approved Zeller's verdict of guilt that Zeller completed his preliminary investigation report. Zeller then sent the report to Romanski, Edwards, and Bitoff for final approval and release.

Zeller's final report is dated 23 October 1989 (in draft form from 5 October). I have released Zeller's report already. I am ready to release it again upon request.

John Bitoff signed out Zeller's final product on 26 October, and Captain Mike Edwards signed on Bitoff's behalf as Bitoff's chief of staff.

Bitoff's claim to higher command is that Zeller's final report should be considered "'raw data,' reflecting specifically the findings, opinions and recommendations of staff JAG Tim Zeller alone, without editing by higher authority."

Bitoff's letter of 26 October, signed out by Edwards, is pure prevarication!

It is a knowing utterance of a false official statement engaged in by Bitoff, Edwards, Romanski, and Zeller.

The 26 October cover letter betrays the underlying 23 October preliminary report, which is also an impressive exercise in mendacity. Zeller's preliminary report stands alone as an illicit writing; Bitoff's cover letter only magnifies its effect.

We now come back to Romanski's memorial.

God bless the man!

All four men know they are practiced liars.

They all know what plan they are hatching.

26 October 1989 was a Thursday.

I am going out on a limb here to gently suggest Paul Romanski and Tim Zeller began work on their **"MEMORANDUM FOR THE RECORD"** the next day, Friday, 27 October 1989.

Romanski's memo is dated and filed on the following Friday, 3 November 1989.

Sometime during the ensuing week, Romanski and Zeller must have compared notes. It was then that Zeller caught a few problems.

Zeller wrote to Romanski, as I paraphrase: "Bitoff's letter signed by Edwards signed out last week on 26 October, forwarding my 23 October report to our three-star officer boss in San Diego. He pointed out clearly that the report was unedited, and the conclusions reached were mine alone."

Well, Zeller knew Bitoff—and Edwards—had lied.

Something Romanski wrote in his chronology for the 5 October entry reveals the lie. Romanski had identified himself as one of my accusers—thus, that writing betrays that Zeller and Romanski were acting as a team. One may read for oneself what Zeller wrote. Zeller's preliminary report is false. Zeller knowingly withheld the identity of Paul Romanski as one of my accusers, and he also withheld the identities of Bitoff and Edwards as my accusers.

Neither Bitoff, nor Zeller, nor Romanski, nor anyone from Bitoff's staff ever contacted the NIS, as required by regulation. No one from Bitoff's staff ever made that attempt.

With God as my witness, all of the charges I eventually faced from my accusers were false.

All of my accusers came from John Bitoff's supply ship staff. Those charges emanated only from Bitoff's staff, thus forming the conditions for a conspiracy: **an evil, unlawful, treacherous, or surreptitious plan formulated in secret by two or more persons; plot, a combination of**

persons for a secret, unlawful, or evil purpose.

Throughout this long telling, none of my shipmates on the *USS MARS* ever accused me of anything.

Zeller's preliminary report was kept secret for well over one year. Remarkably, it was never part of the official court-martial record. It was not available during the Article 32 hearing. It was not available for the April 1990 court-martial.

Instead, Romanski's chronology was kept secret for eleven years, nine months.

Bitoff, Edwards, and Zeller all took part in handpicking my defense counsel: Captain of Marines Kevin Martis "Andy" Anderson.

The same Kevin Anderson who forged my name to the counterfeit confession on 17 July 1990.

The forged confession served its intended purpose: to disabuse anyone outside Bitoff's command structure from taking a closer look at what Bitoff, Edwards, Romanski, Zeller and other supply ship staff officers had done. More information about the other staffers shall follow.

Throughout these complicated and murky machinations, Bitoff and his men all knew exactly what they were doing. And they did not want to get caught in the middle of the evolving treachery.

Bitoff really had nothing to worry about. In this court-martial system as I portray it, men like Bitoff never get caught. .

Yet, that is precisely why I wish to take up your valuable time.

We all need to know how the system can fail.

This is the exercise of unlawful command influence presented, proved, and displayed in a complete manner not ever before possible.

This IS how courts-martial work!

Please consider the documents that I have tracked down and included here for you with care and respect. Trust me, they were not easy to find!

There exists even more incriminating documents still locked in vaults—that is, if they still survive.

The forgery is still being used against me today. This is an active criminal event that extends the life of every other allied criminal act report here and before.

CNO Greenert, Navy TJAG DeRenzi, Kevin Anderson, and so many other officers represent and guard the forgery (one supplied separately) as an original writing bearing my true signature.

They, along with so many others, can still be held to face criminal accountability right now.

To refresh, the rapidly paced timeline, as we are able to gather the documentation, now looks like this:

- **Monday, 2 October 1989:** Zeller returns to staff headquarters in Oakland, CA. Zeller orally briefs Paul Romanski. Romanski orders Zeller to prepare a written report. "Fitzpatrick is a guilty man!" Pointedly, Romanski did not tell Zeller to contact the NIS.

- **Monday, 2 October 1989:** No records exist documenting contact by any naval authority with the NIS regarding Zeller's preliminary investigation.

- **Tuesday, 3 October 1989:** No records exist documenting contact by any naval authority with the NIS regarding Zeller's preliminary investigation.

- **Wednesday, 4 October:** Rear Admiral Bitoff returns to the San Francisco Bay Area. Zeller, only back just two days himself, sends out his first preliminary investigation status report to higher command stating that the NIS had been contacted, but he had been too busy to answer the call! Zeller gives no details regarding who spoke with whom either on 2, 3, or 4 October 1989. No records exist proving any contact with the NIS whatsoever.

- **Thursday, 5 October 1989:** Zeller has the first draft of his investigation report ready to show Romanski in three days' time. It is an extensive report. Romanski and Zeller talk. During Romanski's intercourse with Zeller, Romanski reviews Zeller's written report. Romanski changes (or edits) Zeller's report enough that he joins

Zeller as accuser #2. Zeller and Romanski are aware—both of them together—that they are accusers!

- **Tuesday, 10 October 1989:** Romanski briefs Edwards on the investigation report that, by 10 October, both Zeller and Romanski had authored and edited. In separate writings, and in Zeller's own words, Mike Edwards interacts with Zeller and edits Zeller's investigation enough that Edwards joins Zeller and Romanski as another accuser. Thus, Edwards becomes accuser #3.

- **Wednesday, 11 October 1989:** Nothing to report.

- **Thursday, 12 October 1989:** Zeller briefs Rear Admiral Bitoff, Captain Edwards, and Captain Romanski in Oakland, California, thus confirming the devious secret conduct of the cloaked court-martial. Zeller repeats his pronouncement that "Fitzpatrick is a guilty man!" Bitoff, Edwards, and Romanski extend themselves in their concurrence. Arrangements are immediately commenced for the show—the more public court-martial. Zeller's preliminary investigation report, at this moment, exists only as a DRAFT!

An initial debriefing was delivered to RADM Bitoff and, if recollection is correct, Captain Edwards was present.

The assertion by Lieutenant Commander Fitzpatrick concerning Captain Edwards's alleged acquiescence or approval of the plan for the funeral party was, to the best of Zeller's recollection, not discussed.

Upon return of Lieutenant Zeller to the Combat Logistics Group ONE staff headquarters in Oakland, California, and the subsequent return of Captain Edwards, the subject of Captain Edwards's knowledge of the funeral party funding was raised.

Captain Edwards related that he was unaware of the involvement of Morale Welfare and Recreation funds for the funeral party until well after the incident occurred.

The assertion by Lieutenant Commander Fitzpatrick that Captain Edwards was present at a ship's formation when it [the MWR funding for the funeral party] was discussed was found to be incredible.

Captain Edwards was escorted to the formation to address the crew after the Executive Officer Lieutenant Commander Fitzpatrick had completed his presentation to the crew assembled on the flight deck of the *USS MARS*.

Captain Edwards's presentation concerned the refresher training evolution.

As the investigating officer, Lieutenant Zeller determined that the assertions by Lieutenant Commander Fitzpatrick of Captain Edwards's involvement to be unbelievable.

Notwithstanding this fact, if memory serves, the subsequent decisions in the case were made by RADM Bitoff with recommendations from the investigating officer Lieutenant Zeller and Captain Paul Romanski, Assistant Chief of Staff for Logistics.

The issues discussed involved the disposition of charges based not only on the funeral party incident, but also numerous other irregularities in the *USS MARS* MWR account which were discovered during the investigation.

Rear Admiral John W. Bitoff, then commander of a surface warfare group headquartered in the San Francisco Bay Area, ordered an investigation into the *USS MARS* (AFS-1) MWR account.

Zeller began the draft of his first written report on 2 October 1989. Zeller briefed Admiral Bitoff on 12 October 1989. Yet, Zeller's report was still in draft. Bitoff reviewed Zeller's report wherein Zeller pronounced "guilt." Bitoff approved Zeller's work, thereupon joining in a conclusion regarding "guilt."

Bitoff ordered the commencement of the court-martial process after Bitoff had concluded that "Lieutenant Commander Fitzpatrick is guilty." In making his command decision, Bitoff left no doubt regarding what Bitoff's eventual and final decision was going to be: "Lieutenant Commander Fitzpatrick is guilty."

Attached is Admiral Bitoff's approved report in final form. Zeller's report was signed out and made official under Bitoff's command authority and under Bitoff's unlawful command influence.

Perhaps the reader may wish to focus on item paragraph seven at the bottom of page five of the first PDF record.

Bitoff's endorsement and approval is the second PDF record.

To the point: Bitoff's very first command function in the exercise of his power to discipline and punish subordinates was to establish and pronounce guilt.

Some explanatory notes:

- Respecting your time, only paragraph seven of Zeller's investigation is relevant in support of this submission. Zeller's report to Bitoff is pure fiction. A critical analysis of Zeller's narrative is not appropriate here; however, a critical written analysis is present in the files I sent first to Christine Clarridge.

- Also, please ponder this: Bitoff and Zeller's activity as described here disqualified both men from creating the court-martial. Nonetheless, both men sailed at full speed ahead!

⚓

FALSE STATEMENTS: THE MANY PROBLEMS WITH ZELLER'S INITIAL REPORT

Every bit of Zeller's report is an intentional fraud. For example, Zeller knew, even before embarking aboard the ship, that "Captain Edwards, Chief of Staff to Commander, Combat Logistics Group ONE had been detailed to replace the present Commanding Officer, Captain Michael B. Nordeen, while Captain Nordeen attended his brother's funeral in Greece." The *MARS*, "at the time of the incident, was undergoing refresher training in San Diego, California."

He goes on to write: "During the course of the investigation,

Lieutenant Commander Fitzpatrick asserted that Captain Edwards was well cognizant of the plan to send *USS MARS* crew members to the funeral. There is no recollection that Lieutenant Commander Fitzpatrick asserted Captain Edwards approved of the action. The alleged irregularities involved an expenditure of approximately $10,000.00 to send certain members of the *USS MARS* crew to the funeral of the *USS MARS* CO's brother, Navy Captain William Edward Nordeen, a Naval Attaché who had been murdered by the terrorist group known as 'November 17.'"

Here is how those pre-hearing investigations worked thirty years later: Enter Lieutenant Timothy W. Zeller (a navy O-3), staff judge advocate to John Bitoff and Mike Edwards.[119]

Rear Admiral Bitoff's first reaction was to hastily return any investigatory responsibilities back up the chain of command. Please recall that it was Mike Edwards, Bitoff's XO, who commanded the *USS MARS* in June and July 1988 during those two weeks when MWR funding for the funeral trip was approved and monies were spent. Edwards, as the ship's skipper, was routinely briefed on events as they evolved, approved the funding source once identified and agreed upon, and okayed crew members selected to attend the funeral who were to be absent from the ship during REFTRA while their shipmates were to be strenuously tested.

Additionally, once the sailors were gone, Captain Edwards received a daily written record—a "muster" (attendance) report—wherein all sailors missing from the ship were identified by name, giving the commanding officer an explanation why each man was absent when reasons were known.

Midday every day that a navy ship is underway, the boatswain mate of the watch standing the 08:00–12:00 bridge watch delivers to the commanding officer a group of reports collectively described as the "twelve o'clock reports." Aboard the *USS MARS*, they included a chronometer

---

119 Bitoff, as a one-star rear admiral, held the seventh highest maritime commissioned officer rank—an O-7.

report, a boat report, a fuel and water report, and a muster report.

The bosun is responsible for collecting these various status reports throughout the morning. The officer on deck (OOD) oversees receipt of the written reports, calling about the ship and pressing others on watch who may be tardy in getting their news to the bridge. With all reports in hand, the bosun approaches the skipper (wherever the captain may be on the vessel), salutes, and sounds off: "The officer of the deck sends his respects and reports the hour of twelve o'clock. All chronometers have been wound and compared. Request permission to strike eight bells on time, sir."

Ship's muster is taken early each morning. By the time the muster gets to the commanding officer, it is but a routine formality. This is because if in the morning muster a man turns up missing, the officer of the deck on the bridge is notified instantly and *"MAN OVERBOARD!"* is sounded on the ship's whistle and announced to the crew over the shipboard public address system (the 1MC).

The OOD orders the ship to come around smartly, reverse course, and increase speed in search of the missing shipmate.

Each and every day the *USS MARS* was at sea for REFTRA, from 30 June through 8 July 1988, the boatswain mate of the watch approached Captain Mike Edwards with the day's muster report, announcing "all men present and accounted for." The list consisted of Lieutenant Bradford Ableson, Ensign Darrel Vaughn, Command Master Chief Poasa Fa'aita, First Class Petty Officer Boatswain Mate Mark Middleton, First Class Petty Officer Hospital Corpsman Mark Collins, First Class Petty Officer Personnelman Arnaldo Centeno, Second Class Petty Officer Electrician's Mate Rusty Padojino, and Seaman Radioman Striker A. D. McCree, officially absent with Captain Edwards's permission.

Unconcerned by the deep conflict presented by the admiral's intimate association with Executive Officer Edwards, Bitoff assigned his staff aviation officer, Lieutenant Commander (O-4) Steve Letchworth, to conduct the first exploration of possible financial misconduct. Bitoff was willfully ignorant, as well, of Letchworth's relationship to Edwards. Letchworth, two ranks junior to Edwards in the same organization, was placed in a situation whereupon he would be scrutinizing Edwards's performance in command of the *MARS* while searching for evidence of alleged financial malfeasance. At the same time, back in staff headquarters, Letchworth worked for Edwards, who had input to Letchworth's performance evaluation as the staff's resident pilot.

Matters grew immediately more complex. Bitoff fired Letchworth as preliminary investigation officer before Letchworth ever began work, but not for obvious reasons.

Bitoff was advised of a recent professional dispute between Captain Nordeen's XO in MARS, a surface warfare officer (or ship driver), and Steve Letchworth Group ONE's in-house helicopter pilot. Letchworth, directly after MARS' return from deployment in March 1989, found what he considered a serious deficiency regarding the ship's flight deck (helicopter landing zone) and was pressing to decertify the MARS for flight operations. This is a very big deal for many reasons.

Letchworth was wrong. He was challenged about his determination to decertify the MARS flight deck, and overruled by higher command (i.e., Bitoff's bosses) in San Diego; thus, MARS retained her war fighting critical certification to land and launch helicopters to the embarrassment of Letchworth personally, and the Group ONE staff more generally. The argument had run for weeks and was just beginning to simmer down a little when Bitoff tapped Letchworth as an ostensible "unbiased" finder of fact in ascertaining the proprietary of financial management and operation of the USS MARS MWR account.

It is precisely at this point that Bitoff's true intentions are laid bare and became clear, at least to staffers Bitoff trusted to keep his secrets. Bitoff's staff judge advocate (legal advisor), Tim Zeller, reported years later the selection process Bitoff used to replace Letchworth. As it happened, the pending inquiry had nothing to do with the MARS MWR account.[120]

Bitoff's genuine target was the MARS' executive officer holding the rank of lieutenant commander (O-4). Bitoff initially selected Letchworth, also a lieutenant commander, based only on Letchworth's seniority to XO MARS by date of rank.[121] Once Letchworth was removed, Zeller reports Bitoff was unable to find any Group ONE staff officer who had not had a similar "confrontation" with XO MARS. "These factors were considered to override normal protocol that a junior not investigate a senior."[122] So it came to pass Rear Admiral Bitoff selected his staff judge advocate general, Lieutenant Timothy W. Zeller, to carry out the preliminary inquiry now specifically focused on the single personage of XO USS MARS.

It must be remembered that Zeller was as conflicted by identical professional relationships to Edwards as the junior-senior staff association Letchworth was with Edwards. Clearly, Bitoff did not seem to notice, or mind. Bitoff's selection, then rejection of Letchworth, and subsequent selection of Zeller as investigator occurred in a single day, Friday, September 15, 1989.

Notification to the Naval Investigative Service (NIS) about the investigation ought to have been carried out immediately. However, Rear Admiral Bitoff made the unilateral decision to go it alone, obviously deciding that the deployment of the *USS MARS* made notifying the NIS impractical. Thus, power began to coalesce.

---

120 Tim Zeller's naval message to Navy Captain Glen Gonzales of 16 April 1992.
121 Seniority of one officer over another when both hold the same rank is determined by their relative promotion dates. For instance, an officer promoted to the rank of lieutenant commander today is senior to every other LCDR promoted the next day or later. Officers promoted to the same rank on the same day are reported on a list of names each assigned a line number. So, in this common circumstance, seniority is found through "lineal number" comparisons.
122 Appendix Twelve: Tim Zeller's 16 April 1992 message to Captain Glen Gonzales.

13.
DIRECTING THE COURT-MARTIAL

⚓

DELVING INTO THE DARK DETAILS of the Article 32 or the court-martial proper would be both tedious and unnecessary. Every navy–marine corps JAG legal analysis found in this case, or any subsequent legal discussion or review, is upended and subverted by the simple truth that Bitoff tossed jurisdiction over the side by way of his illicit dealings with Tim Zeller and Kevin Anderson.

In the eyes of military law, when seen clearly, Bitoff was strictly prohibited from convening either military discipline hearing.

For the sake of clarity in understanding, if not for its entertainment value, there are some chillingly remarkable features that must be called out to the gentle reader to provide fully fleshed out and robust dimension to the workings in this military discipline outlaw adventure.

⚓

Tim Zeller was the Integrity and Efficiency investigating officer. Zeller was also one of three accusers alongside Bitoff and Paul Romanski.

Zeller's numerous investigation reports and notes were cloaked. All were kept secret. Contrary to regulations, none of Zeller's writings were made available to prosecution or defense attorneys.

Zeller did not testify during the Article 32 hearing, nor did Zeller testify at the court-martial. Nor did any of my three accusers—all of them officers on the Combat Logistics Group ONE staff—testify at the Article 32 hearing.

Not one of the trio testified at the court-martial featuring a rigged panel of CLG-1 staff officers, either.

⚓

Rear Admiral John W. Bitoff, Navy Attorney (then Lieutenant) Timothy W. Zeller, and Assistant Chief of Staff Paul A. Romanski were my only accusers. None of these men served with me aboard the *USS MARS*, and none of them were my shipmates. All three were staff command officers overseeing the operation of the ship. Bitoff was the staff commander, and it was Bitoff who created the military discipline court I stood before facing accusations leveled solely by Bitoff and Zeller. Bitoff admitted in 1999: "I brought the charges, and I convened the court-martial [...]. LT Timothy W. Zeller was my Staff Judge Advocate regarding this matter. I followed Lieutenant Zeller's advice."[123]

In a separate correspondence, Zeller relates: "In order to ensure that my part in this case was apparent, I signed as the accuser on the charge sheet. This made me ineligible to fulfill the statutory duties of rendering advice to Admiral Bitoff [...] as well as ineligible to review the case for legal sufficiency and errors."[124]

Zeller expounds further: "I was tasked by Rear Admiral Bitoff with conducting an investigation into the MWR expenditures onboard *USS MARS*, said tasking being a result of a directive from Commander Naval Surface Forces, Pacific. My client in this matter, as both investigator and Legal Officer was the Department of the Navy as personified by Rear Admiral John Bitoff."[125]

⚓

---

123 Appendix One: Bitoff's 30 April 1999 statement to Congressman Dicks.
124 Zeller's 8 July 1998 statement to OBA Investigator Tony Blasier
125 Zeller's 8 July 1998 statement to OBA Investigator Tony Blasier.

However, as we have seen, Rear Admiral Bitoff was prohibited from convening both the Article 32 hearing and the court-martial. Bitoff and Zeller were both accusers; one may not convene and accuse at the same time. Zeller declared his cloaked "guilty" verdict, in writing, a full six months before the court-martial in October 1989.[126]

Zeller provided Bitoff with continuous advice and support, notwithstanding black-letter prohibitions against acting simultaneously as investigator, accuser, and advisor to the convening authority on the same case. Black-letter laws are the well-established legal rules that are certain and no longer disputable.

At this point, I shall focus on Bitoff's letter of reprimand that distills the two "crimes" for which I now stand federally convicted:

(1): As evidenced by [the court-martial record], I signed a check out of Morale, Welfare and Recreation funds [...] to pay for a trip to Hawaii. It was determined [during the court-martial] that such a brief was never held and had never been scheduled.

(2): As further evidenced by [the court-martial record], I was warned twice, both before and after the purchase [of electronic equipment], of the impropriety of this purchase.[127]

Worthy of notice is that Zeller initially relied upon a main charge of larceny, accusing me of personally pocketing, stealing, MWR funds used to send a contingent of *USS MARS* personnel to the funeral of Navy Captain William Edward Nordeen, assassinated in Athens, Greece, who was our commanding officer's brother.

⚓

An Article 32 hearing involves an investigatory body, a fact-finding

---

126 Zellers I&E investigation reports 4 October 1989 (and dated 10 and 23 October 1989.) Also see Appendix six the Romanski chronology.
127 Bitoff's letter of reprimand dated 7 June 1990.

body tasked with submitting recommendations to a convening authority regarding the disposition of charges leveled against an accused.

The frequent comparison of military government Article 32 hearings to a civilian grand jury is misplaced. Chronologically, the Article 32 hearing is positioned inside a criminal proceeding in the same place where a civilian grand jury would appear, but that is as far as any proper comparison goes.

⚓

ZELLER'S 1989 THANKSGIVING DAY MEMO

The Article 32 hearing is a function of command.

Zeller and Bitoff ("we") asked for an above average defense counsel for the defense so that the court-martial would be fair and for a military court-martial judge to ensure that the complexity of the case would be appreciated.

On the day after the first day of the Article 32 hearing, Wednesday, 22 November 1989, in the full exercise of unlawful command influence, an anxious, frantic, panicked, and maniacal Zeller came to work on the morning of Thanksgiving Day to author his infamous "ATTORNEY WORK PRODUCT" memo.

Zeller demanded Admiral Bitoff manifest his discretionary command function to fire and replace the independently assigned prosecuting attorney (government counsel, or "GC") before the start of the next day's Article 32 proceeding, a Friday.

Exploiting intelligence that Zeller enjoyed from legally trained subordinates he had sent to watch the hearing (i.e., spies), Zeller alarmingly relayed to Bitoff that Bogoshian not only was incompetent, but that the "GC" also lacked the required dedication of someone who "desires to win" (read: convict). Zeller insisted Bogoshian be cashiered (to be dismissed from the military in disgrace) immediately.[128]

Zeller's Thanksgiving Day memo also chillingly signifies that the Article 32 hearing was just a sideshow, and that its conduct and subsequent result, albeit guaranteed, was an unnecessary step of no real consequence.

As a function of command, and as a clandestine act of supremacy, Bitoff and Zeller had already decided there was going to be a court-martial, no matter what! Assignment of the military discipline hearing officer, the court-martial "judge," was predetermined and arranged.

For goodness' sake, the outcome of the court-martial was therefore predetermined and arranged: Zeller pronounced me guilty in October. My actual "secret" court-martial was convened and conducted on 12 October 1989.[129] Bitoff, Zeller, Edwards, and Romanski were all in attendance. From 12 October onward, everything carried out was subterfuge, mere showcasing for public consumption.

The significance of Zeller's "T-Day" memo is that it revealed that Bitoff and Zeller had a keen stake in the outcome of the court-martial— the goal was to "win" at all costs. Zeller admits to the attorney-client relationship between two of my accusers, and thus exposes them illicitly as deputy prosecutors.

Incidentally, the use of the descriptor of "judge," as Zeller applies it, is repeated advisedly. Calling a court-martial presiding officer a "judge" is like calling Punxsutawney Phil a weatherman. Ha!

⚓

---

128 Appendix Ten: Zeller's infamous 1989 Thanksgiving Day memo.
129 Appendix Six: Romanski/Zeller chronology exchange.

THE *USS MARS* (AFS-1) 1988 ANNUAL MWR FISCAL REPORT – EVIDENCE TAMPERING!

Each command is required to maintain the MWR fund account, and it is to be submitted as part of an annual report sent to higher command at the end each fiscal year. The *MARS* submitted the original copy of its FY-88 MWR report to Commander, Naval Surface Forces, Pacific Fleet in San Diego at the end of 1988. Every cent of money expended from the *MARS* MWR account was painstakingly itemized for all monies paid out for every reason detailed in this narrative.

⚓

Throughout the proceeding, Zeller was intimately involved with blocking, attempting to block, and securing witnesses. Zeller was the primary, and Bitoff was the assistant "second seat" prosecutor. Once charges are referred to an Article 32 investigation, the convening authority must stand clear and allow for the independent detailing of investigators and prosecutors.

Neither the convening authority nor his staff judge advocate are allowed to have a stake in the outcome of a court-martial. Therefore, it is impermissible to "adopt" and advance a position on witnesses.

Lieutenant Brian Feeley was a junior supply officer aboard the *USS MARS* in 1988 and served as the MWR fund treasurer. Brain left the navy after his afloat tour aboard the *MARS* had ended to then matriculate at Cornell University in Ithaca, New York, in pursuit of a postgraduate degree.

Zeller tracked Feeley, now a civilian, down at Cornell in early November 1989 in the hopes of persuading him to voluntarily testify at the Article 32 hearing as a government (i.e., prosecutory) witness. Brian knew exactly what duplicity Zeller was up to and refused to comply. Instinctively, Zeller threatened to criminally charge Brian under military law and to court-martial him should Mr. Feeley persist in his noncooperation.

Zeller did not tell Brian that Bitoff could do no such thing.

Feeling coerced, Brian requested immunity from the promised military prosecution.

Bitoff granted Mr. Feeley's demand, and Brian testified at the hearing by phone. Brian did not appear as a witness for the April 1990 court-martial.

"A convening authority [Bitoff] who grants immunity to a suspect [such as Mr. Brian Feeley][130] who testifies as a witness is disqualified from reviewing the record of [hearing][131] [...] because [the convening authority] might place too much reliance on the testimony of the witness to whom he granted immunity."[132]

BITOFF'S AND ZELLER'S STAR WITNESS: A GHOST! THE PHANTOM!

Lieutenant Commander Doug Dolan was assistant supply officer on the USS MARS during my tour as executive officer.

Doug was Zeller's and Bitoff's star witness.

On page 1 of Zeller's 23 October 1989 Integrity and Efficiency (I&E) report, Zeller catalogs officers whom he had interviewed under "Evidence Examined" (paragraph 5).[133] However, Zeller did not interview Doug Dolan, and yet alarmingly, Zeller cites comments attributed to Dolan in the body of the I&E report (p. 3).

Zeller elaborates on page 8: "The impropriety of the acquisition was pointed out to the Executive Officer at the time of the purchase and afterward by Lieutenant Commander Dolan, the [assistant supply officer]."

---

130 Lt. Zeller's investigation report of 10 October 1989, pp. 5, 17. Zeller wrote that Brian Feeley was suspected of wrongdoing and possibly "subject to subpoena for testimony at trial." Bitoff immunized Feeley, as a suspect, against court-martial in exchange for Feeley's testimony.
131 Byrne, p. 92, citing *United States v. White*, 10 USCMA 63, 27 CMR 137 (1958).
132 Byrne, p. 284, citing *Green v. Widdecke*, 19 USCMA 576, 42 CMR 178 (1970).
133 Appendix Five: Zeller's Integrity and Efficiency report.

Doug Dolan was not called as a witness to the Article 32 hearing in November 1989. At that proceeding, Dolan did not testify.

Doug Dolan was not called as a witness to the April 1990 court-martial. At that proceeding, once again, Dolan did not testify.

There is nothing in the court-martial record regarding Doug Dolan or Zeller's invented testimony claiming Dolan had issued two warnings. Dolan appears only in Zeller's pre-court-martial I&E reports.

Bitoff wrote 228 days later, in my letter of reprimand dated 7 June 1990, "You were warned twice, before and after the purchase, of the impropriety of this purchase." Bitoff cited the court-martial transcript as his source for the statement but did not attribute the words to Doug Dolan, the only person ever identified as having issued the warnings.

Here is the fact of the matter: Doug Dolan never talked to me about pre-deployment equipment purchases at any time. Zeller did not interview Doug.

Doug did not testify at either the Article 32 hearing or the four-day April 1990 court-martial.

Therefore, improbably—and gallingly—throughout these grim proceedings, Bitoff and Zeller used Lieutenant Commander Dolan as a pawn, a phantom, a ghost witness, and all the words attributed to him by Bitoff and Zeller are simply fabrications.

LCD W.D. Dolan
Assistant Supply Officer

Victim of Car Bomb in Athens Buried at Arlington Cemetery

Naval Attache at U.S. Embassy Given Honors

From News Services

A U.S. naval officer killed by a car bomb last week in Athens was buried at Arlington National Cemetery yesterday with full military honors.

Capt. William E. Nordeen, 51, of Centuria, Wis., was interred after a brief graveside service that followed a private ceremony in a cemetery chapel. About 400 people attended, including Nordeen's widow and 12-year-old daughter.

Nordeen, defense and naval attache at the U.S. Embassy in Athens, was killed June 28 by a car bomb detonated by remote control as he drove by. Nordeen's armor-plated sedan was thrown across the street, where it lodged in a steel fence, police said. He was thrown from the car by the blast.

After the reading of the Lord's Prayer, the firing of a 21-gun salute by a Navy honor guard and the playing of taps, the American flag over Nordeen's casket was folded and handed to his widow, Patricia Anne Nordeen.

She briefly clutched it to her chest, then put it in her lap and placed her daughter's hand on it.

Nordeen's mother, Edna Helena Capello, sobbed as she was handed a second flag by an honor guard.

Also attending the funeral were Nordeen's brother, Navy Capt. Michael Brent; Navy Secretary William Ball and Greek Ambassador George D. Papoulis.

A left-wing Greek terrorist group known as November 17 asserted responsibility for the bombing. The group, which is accused of 11 other assassinations in the last 13 years, also threatened to kill more Americans until the United States abandons its four bases in Greece. The two countries are negotiating a new base agreement.

The United States is offering a reward of $500,000 for information about the person or persons who killed Nordeen.

Nordeen's family said he was due to retire in August after 30 years in the Navy.

Victim of Car Bomb in Athens Buried at Arlington Cemetery: Naval Attache at U.S. Embassy G
The Washington Post (1974-Current file); Jul 7, 1988;
ProQuest Historical Newspapers: The Washington Post
pg. A1

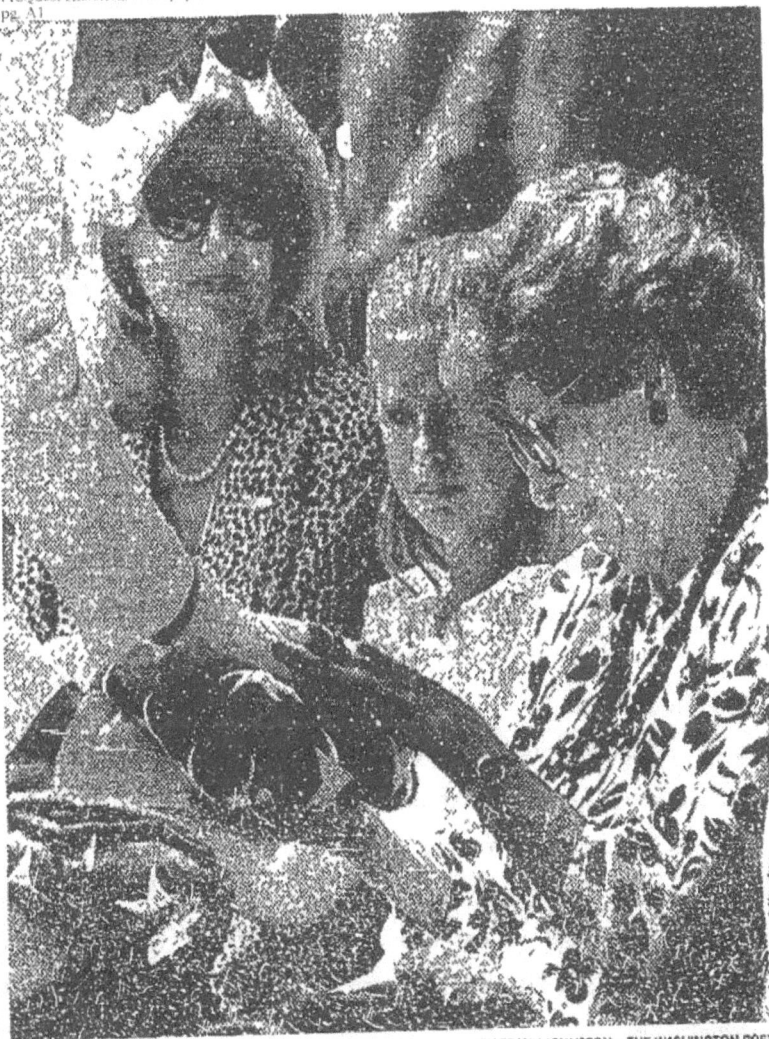

BY FRANK JOHNSTON—THE WASHINGTON POST

REMEMBRANCE

A funeral was held yesterday at Arlington Cemetery for Capt. William E. Nordeen, who was killed by a car bomb last week in Athens. A flag was given to his mother, while his wife and daughter watched. Story on Page A16.

Zeller's I&E investigation report was held secret for years after the court-martial ended.

Finally, Bitoff and Zeller made use of a second ghost witness after the court-martial via statements attributed to myself in the manufactured, forged confession dated 17 July 1990.[134]

⚓

"THE OFT REPEATED FALSEHOOD: AN OFFER OF ADMIRAL'S MAST"

In situations where minor disciplinary offenses are accused, an administrative punishment procedure exists, one known as nonjudicial punishment (NJP)—an action that provides to the convening authority (the designated commanding officer) an alternative to ordering a court-martial. Military jargon uses various descriptors for NJP. The phrase "Captain's Mast," "Admiral's Mast," or "Article 15" are often used in the navy and coast guard. Marines call the hearing "Office Hours."

None of the limited, less severe punishments handed out in a "Mast" hearing carry the same weight as a federal conviction.

After the conclusion of an Article 32 investigation, based upon the nonbinding recommendation of the investigating officer, the convening authority then has the option to offer the accused offender a choice between an Article 32 hearing and a court-martial. This offering, if advanced, is a formal written and witnessed exchange recording the accused's selection.

A collection of documents officially delivered to the accused constitutes a "Mast package." The Mast package is comprised of a catalog of charges (the charge sheet) and papers requiring original witnessed signatures respecting either the accused's acceptance or refusal in demanding a court-martial. Attendant to this step, required endorsement forms must

---

134 The forgery.

memorialize subsequent actions that further establish a chain of custody in the record of the actions taken.

Nothing in the administration of Article 15 is done lightly or cavalierly. There is a formal paper trail for each step.

Inasmuch as Zeller had pronounced my guilt in early October, and my "star chamber" court-martial was conducted on 12 October 1989, Bitoff was not about to offer me the option of an Admiral's Mast. However, Bitoff left it to Zeller to make it appear as though Bitoff had put that option on the table.

As it was his habit to doctor and conceal records to cover higher command when questions were later asked,[135] Zeller constructed an impressive "Potemkin village" facade in his volume of secret memos[136] designed to hide the undesirable fact that Bitoff did not, in fact, offer an Admiral's Mast, while at the same time providing a duck blind for Bitoff and any reviewer to hide behind precisely when questions were asked. Bitoff was adamant in asserting that Article 15 had been offered and that it had been turned down.

Mike Boorda was sworn in as Chief of Naval Operations on 23 April 1994. In a letter to Congressman Dicks on 30 April 1999, Bitoff states: "I had to make a statement on this case in April 1994 when the Chief of Naval Operations [CNO Mike Boorda] directed the Judge Advocate of the Navy [TJAG Rick Grant] to look into the matter following a series of newspaper articles that appeared in your home state of Washington [in April 1994]."

Bitoff's "statement" to CNO Boorda remains concealed; however, it can be taken for granted that Bitoff's report to Boorda in April 1994 was identical to Bitoff's statement to Norm Dicks five years later in April 1999. Therefore, anything CNO Boorda came to know regarding Admiral's Mast came from John Bitoff directly.

---

135 Appendix Four: Zeller's P-4 to Bitoff.
136 Zeller's writings of 11, 16, and 17 January 1990.

As was the premeditated plan, Bitoff relied upon Zeller's tricky witchcraft to make it appear that Bitoff had no other choice but to bring me before a court-martial, defaming me for "irrational behavior" in refusing a nonexistent NJP offer.

Bitoff goes into deep detail about those days when an NJP hearing could have occurred, yet it was all deception.

Until Bitoff's 1999 communication, I did not know about the 1990 meeting Bitoff claims he had with Kevin Anderson (my assigned defense) when, according to Bitoff, Anderson walked in cold on Bitoff unannounced.

No record of the meeting between Bitoff and Anderson is found other than in Bitoff's writings. This merits attention. Zeller would most certainly have memorialized the Anderson-Bitoff meeting in at least an internal memo for file. There's no question that Zeller was in attendance, and Bitoff hints that he was. Today, I maintain that this is one of Zeller's writings that is still being held secret.

Bitoff told Boorda about Anderson's unbelievable "no-knock" entry, which was singularly remarkable, extraordinarily unusual, and just not something that was done.

Boorda knew that there should be a paper trail. Notice of Anderson's offensive arrogance should have triggered an intense search in the records, one focused on discovering evidence respecting Bitoff's make-believe "Admiral's Mast" package.

So, when Navy TJAG Grant's assorted searches came up empty, Boorda knew Bitoff was lying. It is debatable whether Grant, or anyone else, conducted any search for papers they knew did not exist.

Rear Admiral E. F. Tedeschi was the commander of Combat Logistics Group ONE in 1994. Rear Admiral Merrill Ruck relieved Bitoff of command of CLG-1 in 1992, and Ernest Tedeschi relieved Ruck around 1994. Tedeschi, answering a Freedom of Information Act (FOIA) request, betrayed Zeller's subterfuge by stating that no Mast packages were found in CLG-1 files. In fact, there was no record of anything to do with an actual engagement between myself and Bitoff with respect to Article 15. Provided in Tedeschi's answer[137] was Zeller's 16 January 1990 memo for file covering the three-page-long charge sheet Zeller had fashioned as of that same date.

In 1994, Navy TJAG Grant was separately under intense pressure to produce the newly discovered and highly incendiary memo Tim Zeller had penned to Rear Admiral Bitoff on Thanksgiving Day 1989. Grant unlawfully denied Congressman Dicks's FOIA request for Zeller's outlaw missive, and he then lied to US Senator Patty Murray (Washington) twice, saying that Zeller's explosive written demand was not available to him.

Department of Defense Inspector General Vander Schaaf discovered TJAG Grant's malfeasance in mid-1994, filing a criminal referral and naming Grant, along with the Assistant Secretary of the Navy for Manpower and Reserve Affairs Frederick F. Y. Pang. Vander Schaaf also put TJAG Grant on notice.

The criminal instrument bearing the simulation of my name, represented as my confession, was discovered in May 1992 and reported to the NCIS in September 1993. Secretary of the Navy John Dalton's "HOLD CLOSE" and special handling orders were still in effect when the court-martial record was also being guarded as a national secret, whether in the Washington Navy Yard, in an office a few blocks away from Boorda's residence, or possibly sitting on Boorda's desk.

Soon after Mike Boorda was installed as chief of naval operations (CNO), Representative Dicks sent Mike Boorda a written request for Boorda's personal intervention in my case. Boorda took months to respond, apologizing for the delay to Dicks and claiming by way of excuse that Boorda "wanted to make sure [he] had all the facts" before making his reply.

---

137 E. F. Tedeschi's FOIA.

The navy was in a bad way in the summer of 1994 and was unwilling to suffer what was sure to be the cause for additional shame and discredit. Boorda knew Bitoff had lied to him and was deeply aware of what sort of unfortunate consequence lay ahead should word of Bitoff's criminal expedition make it further into the public arena.

So, Boorda, writing in his own hand, was taking Bitoff's dictation when Boorda put pen to paper, inking this statement: "I reviewed this case very carefully. It would not have been a court-martial had the Lieutenant Commander driven it in that direction. NJP was the appropriate level of discipline for dealing with it. Basically, [LCDR Fitzpatrick] presented the convening authority [Rear Admiral Bitoff] 'with a court-martial or nothing' choice and [RADM Bitoff] chose [court-martial]."

Bitoff, Zeller, and Boorda took full advantage of Zeller's protective cover, all three saying that I had refused Bitoff's offer of an Admiral's Mast.

Bitoff, Zeller, and Boorda all knowingly lied. No paper trail of the supposedly offered Admiral's Mast exists.

This played out in precisely the way Zeller and Bitoff had intended, providing Bitoff sufficient cover when "questions were asked later." Altogether, the reader may conclude that this was a very smooth, sophisticated operation.

⚓

THE COURT-MARTIAL PROPER (2–5 APRIL 1990)

First of all, repeating the crucial fact for emphasis, Rear Admiral Bitoff forfeited his jurisdiction over myself and over the alleged criminal act due to Bitoff's self-admitted roles as both accuser and convening authority when he said: "I brought the charges, and I convened the court-martial."[138]

---

138 Appendix One: Bitoff's letter to Congressman Dicks dated 30 April 1999 and his letter to Secretary Danzig dated 4 June 1999.

Furthermore, Bitoff, Zeller, and Romanski predetermined my guilt and conducted a "star chamber" court-martial on 12 October 1989.

⚓

As a matter of military law, the conviction for willful dereliction of duty (Article 92 of the UCMJ) as charged against me was impossible. For the charge to move forward, the government must describe the level of knowledge possessed by the accused at the time the alleged offense occurred. However, Zeller failed to do so in his charging papers.

The military "judge," Navy Captain George Wells, discovered that fact in his examination of Zeller's work, and thus, the pleading as to Specification 1 of Charge I (Article 92) was defective. The pleading required change. Wells wrote: "One cannot be convicted of willfully failing to perform his duties unless he had actual knowledge of the duties." Wells further stated, "[...] the government had not alleged that LCDR Fitzpatrick had actual knowledge."[139]

Nonetheless, the charge went forward undisturbed. Since I was convicted only of the offense that failed to allege actual knowledge, an element of willful dereliction of duty, I was convicted of an innocent act.

This is profoundly significant.

⚓

Bitoff's staff aviation officer, LCDR Steve Letchworth, sparked a particularly violent eruption of the running Hatfield-McCoy feud between the ship's company and CLG-1 staff personnel. In 1988, during *USS MARS* deployment workups, Letchworth decertified the vessel for flight operations—a crucial operational capability—by objecting to the type of nonskid deck coating used, citing it as below required navy aviation operations standards.

---

139 George Wells' Examination of the Court.

His action was nothing but harassment. Higher command settled the dispute in favor of Captain Nordeen, a qualified carrier Centurion[140] navy attack jet pilot. Our skipper's position was that the applied flight deck coating was perfectly fine; however, the matter was not laid to rest until after a series of intense head-to-head arguments with Steve Letchworth.

It was not known until Saturday, 29 September 2001, that Bitoff had initially selected Letchworth as the investigating officer into the *MARS* MWR financial account. Bitoff soon backed off, recognizing Letchworth's still simmering acrimony. He took Letchworth off the case and detailed Zeller as investigating officer instead.[141]

As a function of command, Bitoff detailed many of his immediate staff officers as potential court-martial panel members. Two of those officers, Steve Letchworth and Commander Dave Armstrong, were seated as members on the three-member court-martial panel. Dave Armstrong was Bitoff's staff material officer. The third member was Bitoff's ship repair officer, Robert Orttman.

It was determined that LCDR Letchworth had had an altercation with LCDR Fitzpatrick regarding flight deck issues—and yet, he was then assigned to LCDR Fitzpatrick's court-martial board!

When another outrageous example of Bitoff's command abuses surfaced, in the blatant exercise of unlawful command influence,[142] Deputy Assistant Navy Judge Advocate General (for Criminal Law) Navy JAG Captain K. R. Bryant nevertheless concluded: "[...] there is no factual basis why LCDR Letchworth could not sit as an impartial court-martial member."[143]

Quoting from the relevant section: "Originally the plan was to send the Combat Logistics Group ONE staff officer assigned to Bitoff as the assistant training officer Lieutenant Commander Steve Letchworth,

---

140 Signifying hundreds of carrier flight deck landings.
141 Appendix Twelve: Zeller's *USS Independence* message.
142 Appendix Thirteen and Fourteen: See Bodaly's and Dave's letters.
143 K. R. Bryant letter dated 2 September 2003.

and LCDR Zeller, then a lieutenant, as Lieutenant Zeller was a junior to LCDR Fitzpatrick. [...] Just prior to the detailing, it was decided by Bitoff and Edwards that Lieutenant Zeller would conduct the Integrity and Efficiency investigation alone."[144]

Letchworth did indeed sit as one of my board members during the court-martial. Commander Robert Orttman (senior member) and Dave Armstrong sat as additional jurors. Armstrong was a CLG-1 staff member as well.

All the aforementioned men still alive should be brought before a federal or court-martial trial. Finally, all these men—including Bitoff, Bitoff's sycophants, TJAG Grant, TJAG DeRenzi (and all other navy TJAG involved), all NCIS individuals, and certainty many within the FBI—should be summoned to trial and compelled to tell the whole truth.

Navy Department higher-ups became aware that Bitoff had selected Letchworth as a court-martial panel member even though he was fully knowledgeable of Letchworth's infectious and pernicious influence. They held that dark information under wraps for eleven and a half years. They knew that Bitoff, the convening authority, had exploited a mantle of authority, coloring every aspect of my wrongful prosecution, and Zeller was at Bitoff's ear all the way, offering clandestine counseling. As a function of his command, Bitoff orchestrated my court-martial using his unlawful command influence.

⚓

During the four-day court-martial, Bitoff's staff officers returned to CLG-1 headquarters to openly discuss and brief Staff JAG Zeller on the day's news.

Zeller also collected intelligence from other sources. As he had done during the Article 32 hearing, Zeller ordered legal professionals, two navy

---

144 Appendix Five.

lieutenant JAG officers, to attend the military hearing: Dave Gruber and Valerie Hillicus-Pellegrino. Lieutenant Hillicus-Pellegrino missed part of one day due to personal reasons.

For her dereliction in not following Zeller's command, Zeller dressed her down abusively, leaving left a mark. Hillicus-Pellegrino relates: "When I told Zeller I didn't watch the court-martial all day, Zeller seemed upset. He made it very clear that he expected me to watch the court-martial and report back to him about what had transpired. I was very surprised at how upset Zeller was that I hadn't watched the entire day. [...] I distinctly got the impression that LT Zeller was extremely concerned about the outcome of LCDR Fitzpatrick's case, and Zeller most definitely wanted a finding of guilty."[145]

As Zeller was insanely fixated on ensuring the certain outcome of a guilty verdict on larceny charges, he accused me of personally stealing MWR funds used to send a contingent of *USS MARS* personnel to the funeral of Navy Captain William Edward Nordeen, assassinated brother of our commanding officer Mike Nordeen.

Hillicus-Pellegrino was not the only JAG who disobeyed Zeller's orders. Zeller also ravaged Government Counsel Bogoshian (read: prosecutor) in the 1989 Thanksgiving Day memo described above when Bogoshian failed to heed Zeller's directives concerning witnesses and evidence. For instance, Bogoshian would have no part in Bitoff's threatening and forced participation of civilian Brian Feeley. That was all on Zeller.

Zeller excoriated GC Bogoshian for not endorsing the threat of court-martialing a civilian. Thus, he defamed Matt Bogoshian as unmotivated and inexperienced.

Respecting his experience with and observation of Zeller, Matt Bogoshian wrote:

---

145 Statement of Lt. Pellegrino.

LT Zeller was difficult to work with on this case and I did not enjoy the experience. LT Zeller seemed obsessed with the prosecution of LCDR Fitzpatrick. It was my impression that LT Zeller had a gut feeling, correct or not, that LCDR Fitzpatrick was a bad egg, and LT Zeller was intent on doing everything he could to show that. LT Zeller was a real pit bull on LCDR Fitzpatrick's case.

I remember that LT Zeller was LCDR Fitzpatrick's accuser and that a tremendous number of charges were preferred to the Article 32 Investigation. From my own research as prosecuting attorney, as borne out by the Article 32 Investigation and Special Court-Martial, the majority of charges LT Zeller brought against LCDR Fitzpatrick seemed to have little or no basis in reality, i.e., there was an absence of much if any evidence to support them. After completing my research, I remember thinking that LT Zeller was quite unusual for bringing all the charges he did against LCDR Fitzpatrick.

LT Zeller ensured that I got all the witnesses I needed on LCDR Fitzpatrick's case. It was my impression that a great deal of money was spent for LCDR Fitzpatrick's case.[146]

Throughout these proceedings, Zeller and Bitoff moved people around like chess pieces. Zeller masked the identity of Paul Romanski as an accuser and blocked one person from testifying at the Article 32 hearing. Zeller also tried to avert Captain Edwards from appearing at the hearing, but instead settled for Edwards's perjured testimony. When live witnesses would not suffice, Bitoff and Zeller manufactured phantoms. And so, Bitoff kept Bogoshian onboard knowing that they could achieve their desired outcome regardless, all the while continuing in their crooked masquerade by rigging the court-martial panel.

Throughout these proceedings, Zeller unceasingly badgered Lt.

---

146 Appendix Eleven: Bogoshian's statement.

Bogoshian,[147] so much so that Government Counsel Bogoshian failed to follow Punxsutawney Phil's (Military "Judge" George Wells) instruction to change the pleading against me to actually charge a military criminal offense.

Zeller wanted the surviving Captain Nordeen as well.

It's Bitoff ("B") you see printed at the top.

---

147 Appendix Eleven: Bogoshian's statement.

14.
USS INDIANAPOLIS (CA-35): STILL AT SEA!

Commissioned: 15 November 1932

Flagship, Combined Fleet

Flagship, Fifth Fleet

Awarded 10 Battle Stars

Hit by Kamikaze in Battle for Okinawa

Delivered world's first atomic bombs

to Tinian Island, 26 July 1945

Sunk by Japanese Submarine, I-58

12:14 A.M., 30 July 1945

Ship's Company:

1,197 Navy & Marine personnel

880 survived sinking

316 survived after 5 days in the water

⚓

Inscribed on a WWII-style coffee mug sitting on my desk:

"They that go down to the sea in ships,
that do business in great waters;
These see the works of the Lord,
and His wonders in the deep."

—*Psalm* 107: 23–24

⚓

THE SAGA OF THE *USS Indianapolis* (CA-35) is one of the darkest stories in the history of the US Navy. E. B. Potter was my naval history professor during my plebe year at Annapolis. He did not lecture on the sinking of the *USS Indianapolis,* nor the tragic loss of her crew. The event was barely mentioned in our textbook *Sea Power: A Naval History* (1961), co-authored by Professor Potter and Admiral Chester Nimitz.

Not until my graduation and upon reading Potter's biography of Admiral Nimitz did I discover and come to understand what tragedy had struck the *Indianapolis* crew.

Whereas Eddie Slovik (1920–1945) would have been hard-pressed to have done less in our WWII struggles, Navy Captain Charles Butler McVay of the *USS Indianapolis* could hardly have done more.

Both men, however, were tossed into the capricious, arbitrary, and treacherous shark-infested waters of America's unfair military discipline system. And sadly, each man received a death sentence from the result of their courts-martial.

There exists no better way to paint the US military discipline protocols as a system of selective persecution and prosecution than to stand Captain McVay and Private Slovik next to each other at attention.

THE 20-24[148]

The mid-watch[149] reliefs were up and ready to relieve the 20–24. Those on watch were looking forward to being relieved. Many of their shipmates were sleeping on the weather decks. It was too hot below, as it was July 30, 1945, in the steamy South Pacific.

Most of the crew were on their feet when the two torpedoes slammed into the starboard side forward. Only moments later they were in the water, treading water and clinging to debris, struggling to survive. Clipping along at seventeen knots, her bow blown off, the *USS Indianapolis* (CA-35) became more scoop than ship as she plowed into the water, rolled to starboard, and descended to the bottom of those Pacific waters. Some survivors reported the propeller screws were still turning as the "Indy's" stern went under the waves.

However, the real nightmare began as the attacking Japanese sub (I-58) slithered away.

Today, it is speculated that around 300 of the 1,196-man crew perished outright. More sailors died in the water from injuries moments later. Others died piecemeal from exposure and other causes over the next four days. Soon, sharks arrived, attacked, and killed nearly all the rest.

As the *Indianapolis* sailors and marines met their peril, they were alone and forgotten by the democracy that they had been so instrumental in defending, since, after all, they were the heroes who had delivered the first atomic bomb to Tinian Island just days before. They had done their job. In a record-setting high-speed run (averaging twenty-nine knots!) from San Francisco to Pearl Harbor, the crew of the *USS Indianapolis* gave "Little Boy" and "Fat Man" their next-to-last ride.

Indy's mission was so secret, in fact, that navy brass in the know did not notice when the capital ship failed to arrive in the Philippines for her next assignment, which was to be refresher training for the planned

---

148 The last watch of the day on navy ships: 8:00 p.m. to 12:00 midnight.
149 The first watch of the day: midnight to 4:00 a.m.

invasion of Japan. Neglected, out of the communication loop, and left to die in an unforgiving sea, surrounded by burning oil and circling sharks, the senseless loss of life at war's end through sheer stupidity and incompetence was a scandal the navy could little afford.

And, just then, World War II ended. The two bombings had forced and compelled the Japanese to agree to an unconditional surrender.

Next, public affair flunkies quickly began the insidious navy cover-up by holding a press release announcing the sinking of the *Indianapolis* with heavy loss of life, just one hour after President Truman announced the surrender of Japan.

SCAPEGOATING

Captain Charles Butler McVay III, Naval Academy Class of 1919, was the commanding offer of the *USS Indianapolis*.

Captain McVay was one of the 316 survivors. Therefore, he was perfectly positioned for the role of scapegoat, one sorely needed by navy governors. Convening authority Navy Secretary James Forrestal, in an outlawed action, charged McVay for not taking proper evasive action so as to more effectively counter a torpedo attack.

NavJAG-65-B

OO-McVay, Charles B. 3rd/A17-20(20240)
JAG:B   29 NOV 1945

NAVY DEPARTMENT
Washington, D.C.

To: Captain Thomas J. Ryan, Jr., U.S. Navy,
         Judge Advocate, General Court-Martial,
         Navy Yard, Washington, D. C.

Subject: Charges and specifications in case of Captain
         Charles B. McVay, 3rd, U. S. Navy.

1. The above-named officer will be tried before the
general court-martial of which you are judge advocate, upon
the following charges and specifications. You will notify the
president of the court accordingly, inform the accused of the
date set for his trial, and summon all witnesses, both for the
prosecution and the defense.

CHARGE I

THROUGH NEGLIGENCE SUFFERING A VESSEL OF THE NAVY TO
BE HAZARDED

SPECIFICATION

In that Charles B. McVay, 3rd, captain, U. S. Navy, while
so serving in command on the U.S.S. INDIANAPOLIS, making passage
singly, without escort, from Guam, Marianas Islands, to Leyte,
Philippine Islands, through an area in which enemy submarines
might be encountered, did, during good visibility after moonrise
on 29 July 1945, at or about 10:30 p.m., minus nine and one-
half zone time, neglect and fail to exercise proper care and
attention to the safety of said vessel in that he neglected and
failed, then and thereafter, to cause a zigzag course to be
steered, and he, the said McVay, through said negligence, did
suffer the said U.S.S. INDIANAPOLIS to be hazarded; the United
States then being in a state of war.

CHARGE II

CULPABLE INEFFICIENCY IN THE PERFORMANCE OF DUTY

SPECIFICATION

In that Charles B. McVay, 3rd, captain, U. S. Navy, while
so serving in command of the U.S.S. INDIANAPOLIS, making passage
from Guam, Marianas Islands, to Leyte, Philippine Islands, having
been informed at or about 12:10 a.m., minus nine and one-half
zone time, on 30 July 1945, that said vessel was badly damaged
and in sinking condition, did then and there fail to issue and
see effected such timely orders as were necessary to cause said
vessel to be abandoned, as it was his duty to do, by reason of
which inefficiency many persons on board perished with the sinking
of said vessel; the United States then being in a state of war.

/s/
James Forrestal

No other US officer had ever been similarly tried in WWII for losing his command while under enemy attack![150] Not one! What is more, navy lawyers subpoenaed Japanese Lieutenant Commander Mochitsura Hashimoto, I-58 captain, to testify against the American commander. Hashimoto testified clearly that nothing done by the *Indy* crew would have prevented his attack. How is that true? The submarine's commanding officer carried two manned suicide "kamikaze" torpedoes that he would have launched regardless had Captain McVay somehow defeated attack by more traditional straight running fish.

Navy brass selected as prosecutor an officer who had received the Congressional Medal of Honor, an action done to cagily diminish the status of Captain McVay and properly influence the "court" to achieve a conviction—all actions accomplished with political expediency.

---

150 Kurzman, Dan, *Fatal Voyage*, p. 207.

DESTROYING THE FAMILY

Understandably, the conviction affected Captain McVay and his family in horrendous ways that today we can only mournfully imagine. It was not enough that he had had to suffer the loss of his ship and nearly all of his crew—no, beyond those two large factors, navy officials saw fit to further use the man to satisfy the public appetite for stern accountability. He was appropriately found farther up the chain of command, but also conveniently fixed at the lowest rank possible by way of atonement.

For many grueling years, Captain McVay was hounded by surviving family members who sought to pile heavy blame upon him, guilt wrongly affixed by the military's trial.

Not the least of the suffering was the dishonor, the attainder, and the bills of pains and penalties suffocating Captain McVay.

USS INDIANAPOLIS (CA-35) LEGACY ORGANIZATION: NEWLY DISCOVERED INFORMATION

To this day, Captain McVay's unjust conviction at Navy Secretary Forrestal's court-martial still stands. However, an organization made up of extended families of *USS Indianapolis* crewmen called the Legacy Organization continues to work even to this day to clear the captain's name.

Research for this book began just after my graduation from Annapolis, long before I became a member of the Legacy Organization myself, and it still continues. Unknown, unexamined facts still lurk like phantoms in the navy catacombs; however, one has made its way to daylight that has never before made public until now.

Author, researcher, and historian Dan Kurzman intimates that Fleet Admiral Nimitz signed out an administrative letter of reprimand for Captain McVay,[151] but until now it was not known that Nimitz actually

---

151 Kurzman, Dan, *Fatal Voyage*, p. 207.

USS INDIANAPOLIS (CA-35)

RADM. John William Bitoff

Captain Michael Brent Nordeen

U.S. Navy Capt. William Nordeen
had served as naval attache in
Greece since August 1985.

DR. WALTER FRANCIS FITZPATRICK, II
PROVIDENCE, RI – FEBRUARY-JUNE, 1947

WALTER FRANCIS FITZPATRICK, III
KNOXVILLE, TN - JUNE 18, 2018

RADM. Harold Eric Grant, former Navy Judge Advocate General

On 7 July 1990, the day Kevin Martis Anderson introduced the forgery of my name into the record, I was at the Oaknoll Naval Hospital in Oakland to pick up my wife and new daughter, Cathleen and Angelica, and drive them home. On 7 July 1990, the author took the picture at home.

wrote, issued, and then delivered the reprimand. However, later on, convening authority Forrestal demanded that that letter be withdrawn.

And it was.

Forrestal, with Fleet Admiral Ernest J. King no doubt tickling his ear, made the administrative (noncriminal) reprimand disappear and then criminalized the affair, instead deciding that Captain McVay was responsible for a military crime. At Forrestal's behest, the outlawed court-martial sailed ahead smartly (see above records).

And it turns out that in military affairs, often a pertinent backstory exists, one hidden from public view and full of mean deceit and tricky skullduggery. While earlier serving under Captain McVay's father, Admiral Charles Butler McVay Junior, the older admiral reprimanded King and other officers for having secreted some women aboard ship. Thus, for years King maintained a feeling of bitterness and resentment against the two McVays. Still nursing this grievance from years earlier, Admiral King later did all he could to make sure the younger McVay, Admiral Charles Butler McVay III, was court-martialed, not reprimanded as Admiral Nimitz had suggested.

Thus, a merciless vendetta was to be metered out, one which also could be described as petulant, childish, and vindictive. King, constantly whispering his grouch into Forrestal's ear, obviously had "an axe to grind" and "was going to get McVay." For further study on the details of this salient issue, the reader may consult Richard W. Newcomb's book *Abandon Ship*, published by Henry Holt in 1958.

Pers 3205 lk
20240

Leadway

26 NOV 1945

From: The Chief of Naval Personnel
To: Commander in Chief, U. S. Pacific Fleet

Subj: Captain Charles B. McVAY, III, U.S.Navy - Letter
        of Reprimand Addressed to

1.   The Secretary of the Navy has directed that Captain McVay
be brought to trial by general court martial for his derelic-
tions in connection with the loss of the USS INDIANAPOLIS.

2.   It is therefore suggested that it would be appropriate
for the Commander in Chief, U. S. Pacific Fleet to withdraw
the letter of reprimand addressed to him in this regard.

W. M. Fechteler
Assistant Chief of Naval Personnel

Robin Patterson, head of the Chief of Naval Operations FOIA/PA Program, released the above document to me at 14:28 hours (ET) on 13 May 2021.

The practical effect of Forrestal's Alice in Wonderland outlawed maneuver, one which no one missed, was to declare Captain McVay's guilt before even convening the court-martial.

First the verdict... then the trial. Backwards!

Today, it is not yet established whether Admiral Nimitz actually physically withdrew the written reprimand.

SACRIFICE

Captain Charles Butler McVay III passed at age seventy by his own hand, using his service pistol, on 6 November 1968 on the back porch of his home in Litchfield, Connecticut. His body was discovered by his gardener. For him, World War II, the navy's massive betrayal, the stinging insults and harassment from the public, the endless vitriolic and anonymous phone calls and letters, and finally the navy's cruel cover-up of the whole truth were finally over.

15.
THE AMERICANIZATION OF EMILY

⚓

SINCE ITS INCEPTION ON THE shores of the Pacific Ocean decades ago, Hollywood has been fascinated by the innate and probably unavoidable tensions that exist in the military between the common soldier or sailor and his many superiors. One could create a list of well over one hundred films that deal with this recurring and complex theme, but here are a few with the year of release in parentheses: *Mutiny on the Bounty* (1935, 1962), *Sergeant York* (1941), *From Here to Eternity* (1953), *The Caine Mutiny* (1954), *The Court-Martial of Billy Mitchell* (1955), *Billy Budd, Run Silent, Run Deep* (1958), *The Young Lions* (1958), *The Naked and the Dead* (1958), *Never So Few* (1959), *Up Periscope* (1959), *The Thin Red Line* (1962, 1998), *Billy Budd* (1962), *The Great Escape* (1963), *Zulu* (1964), *Von Ryan's Express* (1965), *The Dirty Dozen* (1967), *Sergeant Ryker* (1968), *The Devil's Brigade* (1968), *Patton* (1970), *Kell's Heroes* (1970), *Too Late the Hero* (1970), *Midway* (1976, 2019), *A Bridge Too Far* (1977), *Apocalypse Now* (1979), *Breaker Morant* (1980), *Gallipoli* (1981), *Heartbreak Ridge* (1986), *A Few Good Men* (1992), and *The Crimson Tide* (1995), among scores of others.

Surely, with just a little judicious digging, it would be possible to expand this list threefold, or more! Even television offers up the inevitable conflicts that take place in the military. One example is Steve McQueen's *Wanted: Dead or Alive* episode 64, entitled "Vendetta" (1960). All of this media presents the simple theme of the innate tension within the military (a factor that has obviously fascinated the American viewer for decades), tension among the common ranks of the soldier and sailor, and tension between the lower and higher ranks. Furthermore, many of them present the picture of the military's own illicit witch hunts, the "stack-the-deck" trials we know and recognize as the courts-martial.

Now, let us concentrate upon one particular film to better convey the point. Using the 6 June 1944 D-Day amphibious invasion as a fictional backdrop, Paddy Chayefsky's brilliant screenplay for the movie *The Americanization of Emily* (1964) offers an apt allegory. The events of the film illustrate the similar lengths to which a malevolent military commander, possessed of image making for "the good of the service," can manufacture or engender "out of thin air" a combat war hero out of a man who is otherwise characterized as craven, reprehensible, despicable, vile, and a self-described coward. In the film's opening scenes, the viewer is soon introduced to that faux "combat hero," Navy "Beach Runner" Lieutenant Commander Charles "Charlie" E. Madison, convincingly played by James Garner.

⚓

As background to better grasp the premise of the movie, it is key to know that on May 19, 1944, President Franklin Roosevelt promoted Navy Under Secretary James V. Forrestal to the cabinet-level position of navy secretary. Forrestal then continued as a combatant in the raging sword fight, one that had been running since November 3, 1943, amongst congressional leaders and senior military officers. That contentious battle raged all across Washington, DC, concerning the proper reorganization and unification of the various branches of the armed forces.

Throughout this lengthy and fractious period, Forrestal was haunted by two crucial factors: firstly, the keen perception that the navy might be stripped of its crucial independence, and secondly, the very real fear that the marine corps might be eliminated altogether. In short, much jockeying for position from all sides of the debate was taking place!

During this debate, army generals focused intently on the United States Marine Corps as a biased competitor for missions, men, and material, targeting the corps with the serious and worrying threat of being "restricted." After the battle for Saipan (June 15–July 9, 1944), Army Lieutenant General Robert C. Richardson described our marines as "beach runners"[152] and robustly exhorted the army's goal of "[...] the permanent restriction of the Marine Corps to a size and functions that would not again resemble those of a regular ground army."[153]

Forrestal took office as secretary of the navy just three weeks before the Allied landings on Normandy beach in France: 6 June 1944, a date we have come to know as D-Day.

⚓

Chayefsky's screenplay for the movie bores penetratingly into the complex issue, profiling the deep-seated concerns navy admirals held that the army was just then intentionally poised to overshadow the navy in the conduct of the imminent D-Day landings.

In the film, Secretary of the Navy Forrestal dispatches Chayefsky's character, two-star Rear Admiral William Jessup (astutely well played by Melvin Douglas), to London as his special assistant to create a navy combat mission—one conveniently folded within the D-Day invasion. A foul deed that, once completed, is designed to nicely set the stage for a report and testimony before a congressional joint committee on military affairs, one analyzing the acute post-war need for a strong maritime navy-marine corps force.

In the Admiral Jessup personage, Chayefsky channels the real-life Secretary Forrestal in portraying a fully distraught figure, someone

152 Caraley, Demetrious. (1966). *The Politics of Military Unification: A Study of Conflict and the Policy Process.* New York: Columbia University Press, p. 68. Also see: p. 68, Smith, Holland M. and Finch, Percy (1948). *Coral and Brass.* New York: Scribner, p. 177.
153 Caraley, p. 66

inflicted by a diseased mind who, in a schizophrenic state, concocts a most bizarre plan. He then orders his flag aide, a navy man, to his death—someone meant to be the first sailor to die on Omaha Beach, thus creating an instant hero who could then be interred in a newly constructed Tomb of the Unknown Sailor.

This desperate action is, of course, a most heinous crime, and the most serious sin that a commander can commit against a subordinate. Please recall again the lines 2 Samuel 11: "And it came to pass, after the year was expired, at the time when kings go forth to battle, that David sent Joab, and his servants with him, and all Israel; and they destroyed the children of Ammon, and besieged Rabbah. But David tarried still at Jerusalem."

In the scant days prior to the invasion, Lieutenant Commander Madison challenges Admiral Jessup and a fellow staff officer in his orders to take part. The admiral chillingly threatens Madison with court-martial if Madison should disrespect his infallibility by disobeying the admiral's orders. Brusquely, Jessup regards Madison's affront as an attack on the essence of military structure, a clear and obvious assault upon the inviolability of command. Jessup authoritatively promises Madison that he will be summarily "brigged," or jailed, should Madison not carry out his assigned duty. Madison's Annapolis graduate staff officer colleague, Lieutenant Commander Paul "Bus" Cummings (played by James Coburn), literally dresses Charlie Madison down, all but cutting off the buttons on Madison's uniform, ripping off Madison's flag aide's aiguillette and service ribbons, and finally excoriating him as "despicable."

On the day of the amphibious assault itself, Lieutenant Commander Madison is threatened with his life, forced to advance on the beach unarmed. Upon retreating, Annapolis graduate, colleague, and Lieutenant Commander Paul "Bus" Cummings, a fellow Jessup staff member, shoots to kill Charlie Madison—however, luckily Bus only wounds Madison, as Madison runs in a zany zigzag fashion all over the beach. Many nearby witnesses believe Madison is killed by a subsequent artillery shell explosion, becoming the first D-Day fatality: a navy man.

Nonetheless, Madison survives!

Madison is then medevacked from France back to London. Bus Cummings quickly confronts Madison with speed akin to summer lightning, getting to Madison before the press does to alert him that he is being lionized as a hero of D-Day and is about to be quickly crowned as a real-life navy hero!

Recognizing that he would be court-martialed on charges for cowardice and misbehavior before the enemy while under fire, Madison is acutely aware that he

1. had embarrassed his country,
2. had dishonored the naval service,
3. had disgraced Admiral Jessup,
4. had humiliated his family, and
5. would likely get thrown in the brig for years, if not shot, for desertion before the enemy.

Madison is staunch in his commitment to destroy the navy's image by exposing Jessup's whole shabby affair to the world for the deranged publicity-grabbing hoax that it was.

In other words, Madison was prepared to make the navy "look really bad" at a time when the navy was striving to justify its reason for continued existence.

Lieutenant Commander Madison was, for a moment, fully prepared to spend a few years in the brig as the price that must be paid for doing the right thing.[154]

However, in the end, Madison caved to the threat of court-martial, whereupon navy mucky-mucks (shortened from the nineteenth-century English term "high muck-a-muck," i.e., self-important officers in unquestioned positions of power and authority) made Lieutenant Commander

---

154 Commander Madison would have been lucky to receive only a few years in a military prison. William Bradford Buie, author of *The Americanization of Emily*, also authored the true story narrative of *The Execution of Eddie Slovik*.

Charlie Madison out as a hero and erected a statue in his honor, awarding him the Navy Cross and Purple Heart.[155]

Unfortunately, as we have seen in real life, the *USS Indianapolis* (CA-35) skipper Captain McVay was not as fortunate. Juxtapose his experience to Lieutenant Commander Madison's fictional one, where Secretary Forrestal was able to contain a massive scandal due to Admiral Jessup's demented hoax; he turned the misadventure around in a foul masquerade to create a false and misleading narrative. In reality, the criminal military acts of a number of senior navy officers, including Secretary Forrestal himself—decisions that left hundreds of sailors to die horrible deaths at sea—were public knowledge. Here was a situation that required both a very rapid cover-up and also a convenient scapegoat, done again for the "good of the service." We may plausibly term this event an early 1945 expression of "cancel culture."

⚓

Forrestal's personal involvement in the conduct of naval operations during WWII is worthy of study if one wishes to understand his off-kilter psyche leading to his outlawed conduct in Captain McVay's November–December 1945 court-martial.

Forrestal, a former WWI naval aviator, lieutenant grade, was physically in the South Pacific periodically from 1942 through 1945. For example, Navy Under Secretary Forrestal was present at the Battle of Kwajalein of the Marshall Islands in January and February of 1944. Additionally, by that time Secretary of the Navy Forrestal, standing on the beachhead, witnessed firsthand the Battle of Iwo Jima in February 1945 and the famous raising of the American Flag. Forrestal then famously exhorted:

---

155 Army Private James Garner, wounded two times during the Korean Conflict, was awarded the Purple Heart with Oak Leaf Cluster (representing two awards). Garner's first acting role was a non-speaking role in the 1954.

"The raising of that Flag on Suribachi means a Marine Corps for the next five hundred years."[156]

Soundly defeating the Japanese at Iwo Jima made the navy and marine corps "look really good." On the contrary, six months later in July and August 1945, the criminal negligence of supreme naval commanders who left the majority of *USS Indianapolis* (CA-35) crewmen in the Philippine Sea, left to die horrible deaths unnecessarily, made the navy "look really bad."

⚓

Screenwriter Chayefsky, had he been so inclined, had an alternate ending to *The Americanization of Emily* with this plot twist: Lieutenant Commander Madison stands his ground, threatening to scuttle the navy's image, and blows the whistle on Rear Admiral Jessup's scheme for the Tomb of the Unknown Sailor, following through on his threat. He is later court-martialed so as to bury the truth and Madison in a single stroke. Secretary of the Navy Forrestal's real-life court-martial of Captain McVay was available as a basis for the script.

The storyline would portray Forrestal's personal and outlawed custom-made criminal accusations, words washed through Forrestal's personally controlled military disciplinary hearing (read: court-martial)—actions carried out to slay Madison and protect Rear Admiral Jessup and the navy from "looking really bad."

Had Chayefsky pursued the alternative ending, he would have positioned himself alongside other literary giants—writers such as Herman Melville, James Jones, and William Bradford Huie—in choosing to condemn America's military discipline system.

---

156 Potter, E.B. *Nimitz*, p. 363.

⚓

In the Captain McVay court-martial, Secretary of the Navy Forrestal, having already tacitly found the beleaguered rear admiral guilty, then worked intently with Navy TJAG Coulouth to customize criminal charges designed to force a court-martial. From its inception, the intention was to create a rigged trial dedicated to depicting only Captain McVay as guilty of an invented military crime.

Then, Forrestal brought the charges and convened the court-martial. Navy Secretary Forrestal personally signed out on Captain McVay's charge sheet.

In a similar and calculated fashion, in the Bitoff-Zeller-Anderson episode I, the defendant, was declared "guilty" as a first act.

Once he had declared the defendant (me) "guilty," Bitoff brought criminal charges and then convened the court-martial. That out-of-order sequence of events is unfair, and it is certainly not how courts-martial are meant to take place. One may pose the telling question: What is the matter with us if this sort of shameful behavior is permitted to take place?

In their sanctification of the carefully planned Bitoff-Zeller-Anderson debacle, TJAGs Hutson and Guter became criminal participants in this less-than-exhaustive catalog of military malfeasance: cruelty and maltreatment, forgery, identify theft, grand larceny, perjury, witness intimidation and tampering, threatening a civilian with court-martial, destruction of evidence, creation of false evidence, false swearing, unlawful command influence, conduct unbecoming an officer, misprision of serious offenses, and attempts towards carrying out outlawed actions.

16.
BUTTON, WHO'S GOT THE BUTTON?

⚓

9 March 1994:

"On February 23, 1993, Lieutenant Commander Fitzpatrick forwarded a request for reconsideration of the action on the March 1993 application, asking in the alternative for a more favorable ruling in this office, certification of the case to the Navy-Marine Corps Court of Military Review, or a new trial. This request was never received in this office."[157]

　　—Two-Star Admiral Harold Eric "Rick" Grant

30 March 1994:

"On February 23, 1993, Lieutenant Commander Fitzpatrick forwarded a request for reconsideration of the action on the March 1993 application, asking in the alternative for a more favorable ruling in this office, certification of the case to the Navy-Marine Corps Court of Military Review, or a new trial. This request was never received in this office."[158]

　　—United States Marine Corps Colonel R. E. Ouellette

---

157 In two separate letters to Congressman Norm Dicks and Senator Slade Gorton.
158 Richard E. Ouellette's letter to Senator Diane Feinstein dated 30 March 1994.

28 April 1994:

*"On 23 February 1993, Lieutenant Commander
Fitzpatrick requested that the Office of the Judge Advocate
General's Corps (in the person of Navy TJAG Rick Grant)
rule more favorably on his request or, in the alternative,
certify his case to the Navy-Marine Corps Court of Military
Review, or order a new trial. That request was not received
and resubmitted by LCDR Fitzpatrick in November
[1993]."*[159]

—Four-Star Admiral Ronald J. Zlatoper

5 May 1994:

*"On February 23, 1993, Lieutenant Commander
Fitzpatrick forwarded a request for reconsideration of the
action on the application on his request for review, asking in
the alternative for a more favorable ruling by myself, Navy
TJAG Grant, certification of the case to the Navy-Marine
Corps Court of Military Review, or a new trial. This request
was never received in my Judge Advocate General's office."*[160]

—Two-Star Admiral Harold Eric "Rick" Grant

10 May 1994:

*"This office of the Navy Judge Advocate General receives
large volumes of incoming mail. If your request for reconsideration was received, it might have been misrouted. We
have no record of its receipt."*[161]

—United States Marine Corps Major R. K. Stutzel

---

159 Zlatoper's memo to Boorda.
160 Navy TJAG Grant's letter to Senator Murray.
161 Major Stutzel's response to postal tracer.

Thus, as it turns out, Rear Admiral Grant did receive the mailing! Duplicity and deviousness are everywhere.

17.
Command Racketeering: Unlawful Command Influence

"Can I call you Court?"
"It sounds so... I don't know—strong and savage."
"Ruthless, sort of..."
"Ah, but imagine if my last name had been Marshall."

⚓

A SCENE BETWEEN COURTNEY MASSENGALE and [insert other character] taken from the book *Once an Eagle*, written by Marine Anton Myrer and published in 1968 by Holt, Rinehart, and Winston. It is a well-regarded standard of military fiction and is required reading at West Point, asking the question: Why do officers start out as Sam Damon and end up becoming Courtney Massengale?

Then, remember 2 Samuel 11 (KJV) and John Paul Roberts's quotes.

⚓

Senior, rather than mid-level, military commanders to this day command the military judiciary. Apologists for the military form of government are quick to point out that changes in the manner in which the military discipline system works are as profound as they are profuse, especially since the enactment of the 1951 Uniform Code of Military Justice (UCMJ).

Yet, the truth of my argument finds its source in this fact: For all of those superficial changes, the role of the commander, the officer in charge, remains completely unchanged from time immemorial!

Uniquely ironic to the birth of our nation is the wholesale adoption, nearly verbatim, of a military form of government that is completely devoid of a judicial process capable of protecting individual citizen soldiers. Ironic, that, because two salient accusations against King George in the Declaration of Independence are: 1) his interference and hindrance of the independence of the judicial process, and 2) his maintenance of a permanent army in time of peace, with many of the troops forcibly housed in private homes.

None other than John Adams tells us that if the military regimen of justice were good enough for the Romans and British, it was good enough for America. And, as we have seen, Adams's Constitutional Convention committee adopted the British Articles of War totidem verbis[162]—in so many words, in these exact words.

⚓

Offering unexpected examples in his book *Man's Search for Meaning*, Dr. Viktor E. Frankl informs us there are two classes of people, and only two: "decent and indecent."[163] Dr. Frankl's experiences occurred at the hands of German military personnel during World War II.

Let us take a look at how Anton Myerer dramatically develops other representations of decent and indecent people in his novel *Once An Eagle*, originally published in 1968. Two characters, both military officers, dominate Myerer's portrayals: Sam Damon and Courtney "Court" Massengale.

---

162 Vallee, James. (1980). *Rocks & Shoals: Naval Discipline in the Age of Fighting Sail*. Annapolis, Maryland: Naval Institute Press, p. 41.
163 Frankl, Viktor E. (1959). *Man's Search for Meaning*. Boston, Massachusetts: Beacon Press, p. 81, 179.

Both are compassionate leaders of men who care deeply about their men, who dig into the foxholes alongside their men, exposing themselves to the same risks, who sincerely care about the health and welfare of their troops both in peacetime and in war. That type of decent leader represents the decent.

On the contrary, those infected and possessed with self-advancement and avarice of rank, who thirst for position and prestige at any cost, who cavalierly and carelessly sacrifice subordinates in their ruthless quest for power, or who use instinctual self-preservation when threatened with exposure for their misdeeds—that type of faux leader is indecent.

Some of these indecent leaders work overtime to shield or hide their past service from prying eyes and avoid any embarrassment and shame.

Then there is the aspect of punishment: to visit retribution on those who have stood up to challenge, or who are perceived as capable of standing up to a challenge again.

Two of Myrer's major plots address the betrayal of trusted friends and associates as they grind up those who have vested power, a mortal sin committed by a senior against subordinates. Then Myrer lays out one of the tools commonly used in a military culture to consummate the betrayal.

⚓

UNLAWFUL COMMAND INFLUENCE

For centuries, commanders have exercised enormous authority in the military's criminal system. Prior to World War I, military commanders were empowered to reverse acquittals and revise sentences that they believed were incomplete or inadequate.[164] Although such unbridled power has been greatly curtailed during this century, commanders continue to legitimately exercise considerable influence over the military justice system, all done ostensibly to ensure unit discipline, morale, and combat readiness.

164 Luther C. West, *They Call It Justice* (New York: Viking Press, 1977), p. 26.

For a moment, let us investigate the example of heroic Jackie Robinson (1919–1972) who first broke baseball's color barrier playing first base for the Brooklyn Dodgers on April 15, 1947. As his earlier court-martial in August of 1944 illustrates (he was charged with insubordination since he would not, upon order, move to the back of a transfer bus), the improper use of command authority to influence the administration of justice has been a persistent fixture in military legal history.

Called the "mortal enemy of military justice,"[165] unlawful command influence refers to a superior's improper interference with a subordinate's duties or discretionary actions within the military legal system. In its most basic terms, improper command influence inhibits the accused's ability to receive a fair trial (Davidson, pp. 128–9). It is a good thing, and quite a lucky one, that Second Lieutenant Robinson was exonerated on the two counts of insubordination by an all-white panel of nine officers.[166]

More commonly described as unlawful command influence, command racketeering is the acceptance and practice of America's extraconstitutional military government while maintaining the illusion that it is actually legal and legitimate. It is what President Andrew Jackson would recognize as the enactment of laws undertaken to institute artificial distinctions, intended to make the rich richer and the potent even more powerful and to make sure that the humble members of military society, junior subordinates, only become weaker and more powerless. Inevitably, it means that the meritocracy (upon which this nation was founded) shall suffer and that the numbers of ingratiating, rear-end-kissing doormats within the military shall increase. This is what William Bradford Huie (1910–1986) aptly described as "a larger State, a smaller soldier."

America's military social order is a pyramid-shaped biosphere where power and influence are simply matters of command tolerance. Powerful

---

165 *United States v. Thomas*, 22 M. J. 388, 393 (C.M.A. 1986), cert. denied, 479 U.S. 1085 (1988).
166 Davidson, M.J. (1999). *A Guide to Military Criminal Law*. Annapolis: Naval Institute Press.

force devolves from the commander in chief onto a group of military nobles populated by only the most senior-ranking officials. Thus, generals and admirals routinely do not go to prison; instead, generals and admirals commonly send weaker subordinates to prison.

In the early days of my research on this book, I would tell others that the practice of unlawful command influence flourishes whilst the law atrophies. However, I have learned that this summary misrepresents the situation. Command influence can be best described as the occasion whereupon commanders of high rank reach into disciplinary hearings to achieve a certain predetermined outcome.

For instance, during World War II, Fleet Admiral Ernest J. King determined that four marines had gotten off too lightly in punishment, and this assessment was adjudged after they had pleaded guilty to their crime. Incensed, and a stickler for harsh and oppressive discipline, King wrote Secretary of the Navy Frank Knox in April 1944, "The sentences awarded in these cases are grossly unmatched to the crime committed." King continued with a request "to advise the court and Judge Advocate of the Navy Department's displeasure."[167]

It would be interesting to determine how King's influence affected the operation of navy courts-martial after the chief of naval operations fired that shot over the bow. Unable to increase the penalty of the marines, members of naval trials while King remained on active duty could not miss the unmistakable message, one transmitted sub rosa ("under the rose," meaning "in secret") throughout squadrons and fleets worldwide. But the exercise of command influence is like advertising—it is simply impossible to measure or gauge when or where it is working well.

On the surface of things, at least ostensibly, command influence was made unlawful in 1951 under the UCMJ, as it was declared a "mortal enemy" to military discipline. However, the practical effect was only to confuse the issue and drive it more deeply undercover. The truth remains

---

167 Thomas B. Buell, p. 346. Buell demurred fully explaining what the Marines did.

that any commander, representing the commander in chief himself, ultimately decides every case regarding the punishment of subordinates. This turns out to be nothing but a one-sided, biased joke.

A commanding officer, deriving his authority from the commander in chief himself, is the final arbiter in every court-martial. All that proceeds their determination (the courts-martial prosecution itself) is nothing more than advisory groups crafting and advancing recommendations up the chain of command to the top.

Thus, command influence is a built-in feature of the military discipline scheme, and as such, it cannot be adjudged "unlawful." It is a classic catch-22. The mortal enemy to US military personnel is not the practice of command influence incorporated into military government by intent and design, but rather the practice of attainder, pains, and penalties. The court-martial of Billy Mitchell is particularly troublesome in its illustration of blind acceptance of idiotic rules.

Airpower advocate Brigadier General William "Billy" Mitchell (1879–1936) would not be as lucky as Second Lieutenant Jackie Robinson a few years later. Mitchell was "attainted" by general court-martial in November 1925 for conduct claimed as "discrediting" and "prejudicial" to good order and discipline. General Mitchell's crime against military virtue was to publicly and clearly denounce the army's and navy's handling of military aviation research and development, with Mitchell accusing both services of incompetency, treason, and criminal negligence. He was most outspoken in his correct comments! The highest-ranking military members ever assembled for an American military disciplinary hearing sat in judgment of Brigadier General Mitchell, including Brigadier General Douglas MacArthur, and the court-martial was personally convened by Commander in Chief Calvin Coolidge.[168]

To some of the higher-ups, General Mitchell's military crime spree

---

168 Davidson, Michael J. (1999). *A Guide to Military Criminal Law*. Annapolis, Md.: Naval Institute Press, pp. 63–64. Davidson cites Burke Davis, *The Billy Mitchell Affair* (New York: Random House, 1967), 240, 247–48.

began when he called a press conference after a series of tragic aviation accidents, culminating in the crash of the airship *Shenandoah* on September 3, 1925, near Caldwell, Ohio.

In his book *A Question of Loyalty* (originally published in 2004 by HarperCollins), *Time* magazine correspondent Douglas Waller tellingly captures the retaliatory mood of senior commanders—among them Commander in Chief Coolidge—as they regarded Mitchell's press statement "incendiary," and close to mutiny. Mitchell had voiced many of these criticisms before, but never in such recklessly harsh language. Mitchell abruptly posited: *Criminals ran the War and Navy Departments? Those who were guilty not just of incompetence, but of treason?* Therefore, by clear extension and obvious implication, Mitchell was charging the president of the United States with high crimes, as President Calvin Coolidge was, after all, the commander in chief of the assembled armed forces.[169]

Illinois Republican Congressman Frank Reid was General Mitchell's chief defense counsel.[170] Congressman Reid used freedom of speech as his constitutionally based defense, while unfortunately failing to recognize the 1925 Articles of War as an act of attainder, pains, and penalties.[171] However, on December 17, 1925, the court of thirteen judges found Colonel Billy Mitchell guilty on all charges, despite the fact that none of the judges had any aviation experience and that all of his allegations about the incompetence of the new Army Air Division had proven to be completely accurate. Colonel Mitchel resigned from the Army Air Service shortly thereafter on February 1, 1926.

Any decision to open up courts-martial must come from the top down. That explains why so few flag officers have ever experienced the

---

169 Waller, Douglas. (2004). *A Question of Loyalty: Gen. Billy Mitchell and the Court-Martial that Gripped the Nation.* New York: Harper Collins, p. 21.

170 Anton Myrer's disgust of the military discipline system is a palpable and recurring subplot in his novel *Once An Eagle.* Myrer makes clear during the Second World War that no person was entitled to be defended by an attorney, and furthermore, no enlisted man ever charged enjoyed vetting of any allegation by other enlisted personnel.

171 Waller, *A Question of Loyalty,* p. 85.

drama of facing discipline. It goes to explain as well why it is well-nigh impossible to advance legitimate criminal complaints against officers in high command. It is assumed that court-martialing admirals and generals is bad for morale and makes the commander in chief "look really bad."

Maintaining military government by attainder, however, only grows more expensive, regardless of how one wishes to measure a nation's treasure. In terms of money, the stakes are much higher in the twenty-first century than ever before, since multibillion-dollar defense contracts represent congressional pork of extraordinary magnitude, contributing to a defense budget that has grown, astoundingly, to half a trillion dollars annually! I must repeat: *annually*!

Alerts to these crucial issues have been sounded before, and yet today they remain ignored.

THE MILITARY LOBBY IS NOW ENERGIZED!

Sensing the swelling political expedient threatening to disrupt the martial order of things, and exploiting an opportunity to maintain internal control, Defense Secretary[172] James Forrestal cobbled together two groups of military men to assume what was widely viewed by elected lawmakers as an unpleasant and unpopular task of making rules for the discipline of US servicemen.

Started anew in 1794, the army and navy each evolved in their own images and traditions as differently from one another other as one community church from the next. Approaches to military discipline were as foreign and diverse in makeup and operation as a tank from a ship. Establishment of the air force in 1947 only compounded already complex and troublesome organizational relationships.

To better understand the issue, we must study the mess that had already been made!

Secretary Forrestal, forced to accept the political reality that reform

---

172 Formerly titled Secretary of War.

and unification of the Articles of War was unavoidable, was coached and encouraged by Senator J. Chandler "Chan" Gurney (chairman of the Senate Armed Services Committee) and adopted a two-pronged approach.

Forrestal gathered an ad hoc coterie of civilian officials "to speak for the armed services" in the preparation of a "uniform code of military justice."[173] Forrestal tapped an undersecretary for each of the three respective services to serve on the newly formed Committee on a Uniform Code of Military Justice (known as the Code Committee). Their task was to craft a UCMJ to apply uniformly to all three services to be submitted to the 81st Congress as a united front from the national military establishment.[174]

But the real heavy lifting was what Forrestal burdened to a subordinate "working group." This group of individuals carried out the difficult task: taking a trinity of service-specific regulations and practices in the scope and operation regarding military discipline and punishment and meld them into one. Populating this group was at least one military representative selected by each of the three undersecretaries named to the Code Committee plus a number of lawyers and civilian researchers within Forrestal's office. Felix Larkin, a full-time staffer to Secretary Forrestal, was named chairman and executive secretary.[175]

Responsive to the clear "implication and impact"[176] of the trend to civilianize military government, Chairman Larkin emphasized the desirability of "keeping [courts-martial cases facing habeas corpus challenges] out of the federal courts." This could only happen if military justice actually represented "a system which would perhaps bend over a little backwards to fulfill" basic requirements of due process. If the military establishment was unable to draft such a procedure that worked, one would ultimately be created for it.

---

173 My emphasis. The word "uniform," in this context, means sameness.
174 Lurie, p. 90.
175 Lurie, p. 90.
176 Lurie, p. 97.

Indeed, it can be argued that the decision to establish dual committees, both dominated by civilians, to construct a uniform code was recognition that the military had been unable to impose on itself the type of military justice reforms sought by the legal order.[177]

The draconian measures laid out in the Articles of War, *Rocks and Shoals*, or the UCMJ, no matter what it may be called, are only justified—if ever—when America is in extremis.[178] This only refers to a time of war or other national danger in which the life of the country is at stake. The British Mutiny Act actually included operative dates, cradle to grave, for that harsh legislation.

Articles of War are temporary and drastic, and they ought not to be tolerated a moment longer than the general public believes necessary. Thus, Articles of War have a beginning and an end.

Putting Articles of War into operation is an extraordinary exercise of presidential war-making power, one absolutely necessary to the effective defense of America. Vesting power to conduct war in the commander in chief, however, is not a blanket allowance for its permanent application. Too late—once the country is under active attack, federal legislators looking ahead wisely empowered the president with the flexibility to instantly respond should that need arise. In our age, the UCMJ is that empowerment—that lays *dormant* until such time as Congress declares war. Actual war, once properly declared, is the only justification for an unconstitutional government supplanting an otherwise constitutional construction.

When the "sleeping giant" is awakened, the commander in chief engages the enemy in one of his first responses by enlivening the Articles of War (think UCMJ) through the vehicle of executive order.

At least that is the intended concept.

---

177 Lurie, p. 97.
178 The "in extremis" condition is defined, in part, by the impossibility that regular federal courts are able to try cases. In short: regular federal courts are closed.

AMERICA'S ARTICLES OF WAR, THE UNIFORM CODE OF MILITARY JUSTICE

The Uniform Code of Military Justice (the Code, or UCMJ) is a *congressional enactment* making a law the military government with its courts-martial system. Before it was renamed in 1950, the UCMJ was known as the Articles of War for the Army and the Rules for the Government of the Navy (nicknamed "*Rocks and Shoals*"). Each numbered section in both of these sequential listings was described as an "article" (synonymous with paragraph, section, or segment). Congress, through the creation of the Manual of the Courts-Martial contained in the UCMJ, authorizes the commander in chief to unilaterally and specifically determine exactly how the courts-martial system is to function.

The Code covers, in general, subjects including the extent and control commanding officers have over subordinates, the types of courts-martial, the composition of those disciplinary hearings, qualification requirements for many of the participants, the role and punishment power of the convening authority, and specific punishments and limitations on those punishments.

THE MANUAL FOR COURTS-MARTIAL

The Manual for Courts-Martial is like a light switch for use only by America's chief executive. Turning it on activates the UCMJ. Turn the UCMJ off and the military courts-martial system goes dark and dormant, removing authority for any person to open courts-martial. Then, only constitutionally recognized federal courts are left to prosecute American servicemen for any and all criminal allegations. The "other" military government is subordinated to the US Constitution's Article III.

Commander in Chief Harry Truman turned the switch on February 8, 1951, under Executive Order 10214. The Manual went into force and effect, with the Uniform Code of Military Justice, on May 31 that same year.

It is important to note that since that date, no commander in chief has ever turned the switch off.

Thus, the commander in chief is responsible directly for the Manual for Courts-Martial. Taken together, the Manual for Courts-Martial and the UCMJ are source authorities, working documents in the scope and operation of the United States discipline and punishment system.

Each of the two documents undergoes annual review—a performance audit, if you will—by each of two separate groups of military advisors to the commander in chief.[179]

The more senior Joint Service Committee reviews the Manual for Courts-Martial and also screens the Code Committee's suggested changes to the UCMJ. Review, modification, or changes to the Manual for Courts-Martial are command functions, like punishment—powers vested solely in the commander in chief. Congress plays no part in review or change to the Manual for Courts-Martial.

Changes or modifications to the UCMJ are legislative functions— literally acts of Congress. Still, new legislative language or suggested revisions to the UCMJ, work products of the military lobby headquartered in the Pentagon, must be vetted first by the commander in chief before those documents are released and recommended to Congress for passage.

I, for one, do not believe any court-martial action is taken on behalf of a single representative or congressman. Unfortunately, the entire process is conducted entirely in secret.

---

179 Joint Service Committee on Military: https://jcs.defense.gov. *The Uniform Code of Military Justice* and *Manual for Courts-Martial*. Retrieved from: https://mdwhome.mdw.army. mil/docs/media-documents/ucmj.pdf.

18.
MR. SMITH GOES TO WASHINGTON!

"Robbin' kids of nickels and dimes."

⚓

IN THE FRANK CAPRA'S 1939 film *Mr. Smith Goes to Washington*, US Senator Jefferson Smith (played by Jimmy Stewart) is betrayed by a man whom he had long admired from his youth. The traitorous Brutus of this tale accuses Senator Smith, under oath, of "robbin' kids of nickels and dimes," relying upon a forgery of Smith's name to a land deed.

That film proves once again that forgery is a very sharp knife, indeed.

ANALYSIS: FORGERY! LOOTENANT CULPEPPER'S DEFENSE

Army TJAG Samuel Tilden Ansell's 1919 famous rage against the military's discipline system correctly observed that for men such as Bitoff, Zeller, and Anderson, it "is axiomatic [...] that they want to get at the fact [no matter how] for the sake of discipline. [...] There is no better witness against a man than himself."[180] The sublime coincidence of TJAG Ansell published polemic coming out of Brian Feeley's future postgraduate home, Cornell, is a fact which must not go unobserved, lest the irony be lost upon John Bitoff and Tim Zeller.

Ansell's remarks are an attempt to force "incompetent confessions" as "the basis of military third-degree methods,"[181] such as those Gregory Vistica vividly describes and which were used against navy and marine

---

180 Ansell, pp. 12, 13.
181 Ansell, p. 13.

corps aviators during the infamous Tailhook investigations of the early 1990s.[182]

$$\text{⚓}$$

"Confessions" are often obtained by preying upon the hopes or fears of those accused, depriving them of the freedom of will or self-control necessary to make a concise voluntary statement. In the military, statements may be fiercely extracted from witnesses and those accused by various threats of violence or gifts of promise, however slight. These illegal acts of coercion may be mental as well as physical.

I call it the "Lootenant" Culpepper's Defense since it is no defense at all, since—and this fact is key—all accused military men are guilty at the outset! In America's unfair and coercive military courts-martial system, "it is absolutely legally impossible to get [...] an acquittal."

Those familiar with James Jones's novel *From Here to Eternity* (1953) know that Culpepper's client, Army Private Robert E. Lee Prewitt, played by Montgomery Clift, was an innocent man... not that it made much difference.

I have embraced the teachings of W. Edward Deming, American engineer and statistician (1990–1993), for many years, especially his emphasis upon the business practice regarding trend analysis. Watching trends is also aggressively advocated as a best practice to be followed in the engineering disciplines.

It is significant, then, to recognize that in the study of military discipline, the anomaly (the outlying discrepant practice) is when an advocate employed to defend a military member actually put up a fight protesting their client's innocence.

Service members are forced, under extreme duress, to speak against themselves. The same is true for uncounted other military personnel

---

182 Vistica, *Fall From Glory: The Men Who Sank the U.S. Navy.*

abducted and locked away in Defense Department dungeons. Forced confessions and the subornation of evidence and testimony is a routine and regular occurrence, one institutionalized within the defense establishment's system of discipline. Civilian criminal convictions in our time are consistent with our understanding of the human condition: that even when seated, a jury does not guarantee a defendant, even one falsely accused, safe refuge or full legal protection. Author and attorney John Grisham arouses deep concern in his nonfiction work entitled *The Innocent Man* (published in 2006 by Doubleday) that the combination of lying accusers and politically motivated prosecutors and politicians are not trustworthy arbiters when evidence is so easily manufactured, contradictory, defective, deficient, or altogether lacking. The scheme of "judicial review" through an appeals process is intended as a corrective whenever jury manipulation or other prosecutorial mischief is uncovered. However, as we have seen, in the military discipline scheme, proper judicial review is routinely tossed over the side.

Simply put, I hope that the reader at this late station will concur with me that the military discipline system is purely a function of command. There is nothing "judicial" about the Uniform Code of Military Justice.

Because criminals in command and command racketeers know there is no review process other than their own, military discipline governors and practitioners grasp that the easiest way to solve and prosecute a crime is to compel a quick confession. Former US Army Judge Advocate General (TJAG) Samuel T. Ansell expressed the concept more forcefully, repeating that "there is no better witness against a man than himself" (Cornell Law Quarterly, November 1919). Army TJAG Ansell warned Americans fourscore years ago that the prosecutorial mischief and tyrannous interrogation methods (instruments that Grisham condemns in civilian practice today, as should we all) have been and continue to be aggressively practiced as standard operating procedures in America's defense establishment throughout—unconstitutionally—all of US history!

Ansell writes:

While the military mind is intolerant of protective principles and of rules governing a [civilian] trial, it is particularly so to the rules of evidence. The professional officers of our [military] in great numbers believed [...] that the business of courts-martial is not to discuss law, but to get at the truth by all the means in its power. [...] We [...] want to get at the fact [no matter how] for the sake of discipline. There is no better witness against a man than himself. That statement is axiomatic among professional officers. They [*the admirals or generals—America's "Flag officers"*] will hear of no qualifications nor can they see evil consequences of the generous application of what is so good. It is the basis of military third-degree methods.

Describing those third-degree methods, we soon discover how special agents from each of the service detective agencies conduct themselves in the most egregious and oppressive manner: the Naval Criminal Investigative Service (NCIS) for the navy and marine corps, the Criminal Investigative Service (CID) in the army, the Air Force Office of Special Investigations (AFOSI), the Coast Guard Investigative Service (CGIS), and then the Defense Criminal Investigative Service (DCIS).

Special agents from the federal goon squads named above extract statements as standard operating procedure (SOP). They utilize psychological tactics of coercion that feature the length of the interrogations, the time of day (night or day, or just after coming in from an extended or grueling battlefield patrol/foot patrol), and exploiting possible flaws in the psychological makeup and military training of the person under scrutiny.

Compounding this, military detectives are not obliged to honor the request of an accused for an attorney. In many instances, such as in the situations of field interrogations carried out in combat theaters overseas, an advocate for the accused just is not available. But no worries—none is required!

Language just used in describing these NCIS/CID/CGIS/DCIS/AFOSI coercive, third-degree tactics would set off alarms for most decent attorneys, inasmuch as it is taken directly from a body of Supreme Court rulings—ones condemning those same interrogation methods.

Defense Department agents also remain unconcerned about any legal constraints otherwise describing them as outlaws. Astoundingly, interrogations are neither audio nor video recorded, and even if they were, these renegade agents could still (and do) operate freely. No other federal law enforcement agency will effect an arrest of a Department of Defense detective, even given clear and obvious evidence of criminal conduct.

And, once in the discipline hearing rooms, courts-martial are not subject to a judicial review of any kind. Therefore, rogue JAGs do not have to concern themselves with the case they have to present to a jury, because there are no juries. Moreover, JAGs find themselves as unconstrained in their criminal conduct as the rogue Department of Defense detectives. There exists no such animal as an "appeal" for servicemen or their families; there is simply no opportunity to seek a later remedy or relief. JAGs are not even subject to discipline from their respective state bar associations. (Dear reader, please note: courts-martial are not proper courts.)

For a moment, let us reflect: What are the motives behind inventing a case against an airman, marine, soldier, or sailor? A very long list comes to mind! A personal vendetta, revenge for an ancient perceived slight, simple petulance, childish vindictiveness, ad infinitum. For the moment, it serves us well to paraphrase a comment made by Admiralty Lawyer and maritime author William James (1881–1973) so far as the character of any court-martial is concerned...

"I want my name back!" In *Mr. Smith Goes to Washington*, a forgery is used to destroy the name of Jefferson Smith.[183] And, as we have seen, the same thing happened to me.

Rear Admiral John W. Bitoff, as the convening authority to my court-martial, placed his Staff Judge Advocate Timothy W. Zeller and assigned Defense Counsel Kevin Martis "Andy" Anderson under orders to rig the enterprise. As it is here recorded for all time, this threesome, working in very close coordination, manufactured false papers, secreted papers, and destroyed papers to secure my certain conviction.

As a function of command and an act of supremacy, Bitoff ordered the manufacturing of a forged "confession" using my name. The bogus statement was part and parcel of Bitoff's insurance policy to mask his misdeeds in the conduct of the court-martial in the event it came under serious later review. Again, there is no such thing as an "appeal" in the military criminal justice process—just review by senior officers. Bitoff ordered Anderson to create the purported admission against self-interest, and then directed him to write my name in an attempted simulation.

Anderson complied with Bitoff's command, and then turned the counterfeit document onto Bitoff via Staff JAG Zeller.

I was in the Kitsap County Sheriff's Office in Washington State making a report about local corruption. The desk sergeant, overhearing my words, came out into the lobby area. She said to me, "You know, I've seen a report with your name on it."

"What are you talking about?" I asked.

The sergeant said that she could not tell me more. When I asked for a copy of the report, she explained a copy was available from the Port Orchard Police Department.

I drove directly from the sheriff's office to the Port Orchard Police Department building a few blocks away. The female receptionist refused to produce a copy of the report on demand, prompting my immediate written public records request.

---

183 *Mr. Smith Goes to Washington*. Frank Capra, Director. Starred Jimmy Steward.

I obtained a copy of a Port Orchard police report on 17 September 2004. Anderson had filed a false police report naming myself about two years before. Anderson's complaint was that I was harassing his family.

Detective Beth Deatherage never followed up on Anderson's perjured accusation. The detective never questioned me, either, which is why I did not learn about Anderson's prevarication at the time.

Instantly, I attempted to file a criminal complaint against Anderson for filing a false criminal complaint. The cops laughed up their sleeves at me, saying, "Well, you know we can't touch this! Fergettabout it!" I had supposedly waited too long to advance the complaint—never mind its extant discovery—and beyond that, they were not going to get involved in the personal contest between myself and Anderson.

Detective Beth Deatherage took Anderson's sworn statement. Anderson was an assistant district attorney under Russel Hauge at the time. Hauge was present as an eye-and-ear witness to Anderson's utterances.

When asked, Anderson told Detective Deatherage that I was allegedly threatening Anderson's family because of a grudge I held against him, stemming from Anderson's participation in my court-martial as assigned defense counsel. Anderson told Deatherage that I had named Anderson as the person who had forged my name to a confession on one of the discipline hearing's writings. Anderson went on to admit, in an ungraded moment, that he was, in fact, the author to the document in question.

NCIS special agents interviewed Anderson in January 1998, five years before Deatherage interviewed Anderson in 2003. Anderson lied to the NCIS agents, telling them he had no recall of the counterfeit confession.

Navy Secretary John H. Dalton wrote in June 1998:

Allegations which brought into question Lieutenant Commander Zeller's suitability for promotion to Commander have been resolved. Investigation into this matter by the Navy Criminal Investigative Service (NCIS) and a complete review of the case by the Navy Judge Advocate General have both determined there was no misconduct by Lieutenant Commander Zeller and the alleged misconduct is determined to be unsubstantiated.

Zeller's Oklahoma State Bar (OBA), meanwhile, had taken up a serious investigation into Zeller's criminal mischief. Panicked, Zeller commissioned Mr. Fredrick Dudink, a professional document examiner, to inspect the questioned signature. Knowing the counterfeit instrument was forged, Zeller's intent was twofold: (1) to show Zeller was not the forger, and (2) to be able to redirect attention away from himself and Kevin Anderson, asserting I had signed the forgery myself.

In defense, Zeller turned to Navy JAG Captain Glen Gonzalez, who was quick to come to Zeller's side. Everywhere I turned, a door was slammed in my face.

One question arises: Is this the best we can do in this country?

19.
IN THEIR OWN WORDS!

"It may suit Americans to invent any falsehood, no mattered how barefaced, to foist a valiant character on themselves."

—William James, Admiralty Lawyer[184]
(1881–1973)

⚓

THE SEISMIC IMPACT BITOFF'S CRIMINAL expedition had upon the military establishment can be readily gauged by the words of senior military and civilian officials who then fervently worked to erase the history of Bitoff's outlaw industry.

The lengths to which military establishment officials will go to maintain their separate government as autonomous is exposed in their own words. As we have witnessed, perjury, witness intimidation, coercion, interference, fabricating and tampering with physical evidence, forgery, and conspiracy have all played a role throughout this criminal expedition.

Yet, itemizing the playlist list of this criminal campaign does not fully express the enormity of the cover-up. With willful disregard for the Constitution, advocates of Colonel Winthrop embrace the mistaken belief that military government is a better form of government. However, for the next decade and more, using all possible avenues open to me, I continued to appeal my conviction at the court-martial. In the fairness vacuum created, and absent any sort of nominal civilian judicial review, here are the results of those lengthy appeals.

---

184 Toll, William T. (2006). *Six Frigates: The Epic History of the Foundation of the U.S. Navy.* New York, New York: W.C.W., Norton & Company, p. 461.

IN THEIR OWN WORDS

Commander, Combat Logistics Group (ONE):
"Since I don't believe in keeping a file to cover this office when decisions are later questioned, there is no copy of this letter in my files on my computer."

—Timothy W. Zeller
Lieutenant, JAGC, US Navy
Staff JAG to RADM Bitoff
In a "PERSONAL FOR" memo dated
11 April 1990 by allied papers

⚓

"In response to your complaints [...] I have determined that the actions of the general court-martial authority [...] are correct, and I approve them."

—Barbara Spyridon Pope
Assistant Secretary of the Navy
(Manpower and Reserve Affairs)
In a letter dated 6 February 1992

⚓

"LT Zeller was difficult to work with on this case. [...] LT Zeller seemed obsessed with the prosecution of LCDR Fitzpatrick. [...] Zeller had a gut feeling [...] that LCDR Fitzpatrick was a bad egg, and Zeller was intent on doing everything he could to show that. [...] LT Zeller was a real pit bull on LCDR Fitzpatrick's case [...]. There was an absence of much if any evidence to support [the charges Zeller leveled against Fitzpatrick]. [...] It was quite unusual [for Zeller] to bring all the charges he did. [...] Zeller was Fitzpatrick's accuser."

—Matthew Bogoshian
Lieutenant, JAGC, US Navy
Government Counsel (prosecutor)
Sworn statement dated 8 April 1992

⚓

"[Lieutenant Zeller's] status as accuser disqualified him from any involvement in the case as staff judge advocate to the convening authority [Rear Admiral Bitoff] either before or after the case or direct involvement in the prosecution of the case."

—Harold Eric "Rick" Grant
Rear Admiral, JAGC, Navy TJAG US
Navy Judge Advocate General
In a letter to the US Senator Patty Murray (D-WA) dated 5 May 1999

⚓

"As the accuser in LCDR Fitzpatrick's case, however, LT Zeller was disqualified from providing the convening authority [Rear Admiral Bitoff] with formal advice on the case. There is no evidence in the record LT Zeller violated this prohibition."

—C. M. LeGrand
Rear Admiral, JAGC, US Navy
Acting Judge Advocate General
In a letter to US Senator Patty Murray (D-WA), dated 9 June 1994

⚓

"[...] article 66, UCMJ does not provide for review [of your court-martial] by either the Navy-Marine Corps Court of Military Review or by the Court of Military Appeals. Since your case has already been reviewed by the [Navy Judge Advocate General Harold Eric "Rick" Grant], pursuant to Article 69, your appellate rights have been exhausted."

—Alice B. Lustre, Lieutenant, JAGC, USNR
Appellate Defense Counsel
In a letter dated 21 July 1994

⚓

"Contrary to your assertions, your record of trial was carefully and objectively reviewed by several officers on various occasions. Indeed, my reviews of your case revealed that you were treated in accordance with the law, and that your claim of unlawful command influence is unfounded."

—Harold Eric "Rick" Grant
Rear Admiral, JAGC, US Navy
Navy "TJAG" Judge Advocate General
In a letter dated 17 October 1994

⚓

"[...] if you can prove the forgery, it totally supports [Fitzpatrick's] 10 years' worth of contentions and makes the [Navy and Marine Corps] look really bad."

—Ernie Simon
Naval Criminal Investigative Service (NCIS)
Assistant Director for Criminal Investigations
In an internal NCIS memo dated 5 September 1997

This determination also put every navy TJAG, the FBI, and congressman in a negative light as well. Some people are still alive who can be prosecuted for this crime.

⚓

"Allegations which brought into question LCDR Zeller's suitability for promotion to Commander have been resolved. Investigation into this matter by the Navy Criminal Investigative Service (NCIS) and a complete review of the case by the Navy Judge Advocate General have both determined there was no misconduct by LCDR Zeller and the alleged misconduct is determined to be unsubstantiated."

—John Dalton, Secretary of the Navy
In a memorandum to Frederick Pang, Assistant Secretary of the Navy
(Force Management), dated 11 June 1998

⚓

"The questioned document examination results revealed 'no conclusion can be reached as to whether or not Mr. Fitzpatrick or Mr. Zeller authored the questioned signature. [...] This is due in part to the poor line quality in the questioned signature. [...] This investigation was closed on 3 February 1998 due to insufficient evidence to prove the allegation of forgery.'"

—P. Cole Hanner
Naval Criminal Investigative Service (NCIS)
Deputy Assistant Director for Criminal Investigations
In a letter to US Senator Slade Gorton (WA) dated 30 March 1998

⚓

"In the fall of 1989, I was tasked with conducting an investigation into the MWR expenditures onboard *USS MARS*, said tasking being a result of a directive from Commander, Naval Surface Force, Pacific Fleet. My client in this matter, as both an investigator and Legal Officer, was the Department of the Navy as personified by Rear Admiral John Bitoff, the Commander, Combat Logistics Group (ONE) [...].

"[...] It is most likely that [LCDR Fitzpatrick] signed the [17 July 1990 letter] himself [...].

"[...] In the past several years, I have been investigated by Commander Naval Surface Force, Pacific, the Judge Advocate General of the Navy, the professional Responsibility Rules Counsel of the Judge Advocate General Corps (twice), the Naval Criminal Investigative Service and myriad other persons [...]."

—Timothy W. Zeller
Commander, JAGC, US Navy (Ret.)
In a letter to the Oklahoma Bar Association dated 8 July 1998

⚓

"My staff spoke with Special Agent Nance, and I personally spoke with Rear Admiral Guter (my deputy TJAG) [...]. [...] Repeated investigations have failed to demonstrate any substance to your many and long-standing claims of improper treatment."

—John D. Hutson,
Rear Admiral, JAG, US Navy (TJAG)
Navy Judge Advocate General
In a letter dated 20 November 1998

⚓

"I brought the charges, and I convened the court-martial. [...] I believe the only option open to you to bring some humane closure to this tragedy, is to convince the Navy to review this case again in light of the troubling allegations Commander Fitzpatrick has advanced."

—John W. Bitoff
Rear Admiral, US Navy Retired (One Star)
Commander, Combat Logistics Group ONE
In a letter to Rep. Norm Dicks (D-WA) dated 30 April 1999
In a letter to Navy Secretary Richard Danzig dated 4 June 1999

⚓

"I am bothered by LCDR Fitzpatrick's recent allegations of misconduct by his defense counsel [Marine Captain Kevin M. 'Andy' Anderson] and the doubts as to the validity of Fitzpatrick's signature on the Response to Letter of Reprimand. I would ask that the appropriate authorities look into these allegations and determine their veracity."

—John W. Bitoff
Rear Admiral, US Navy (Retired)
Former Commander, CLG
In a letter to Navy Secretary Richard Danzig dated 4 June 1999

⚓

"Calling a court-martial presiding officer a 'judge'
is like calling Punxsutawney Phil a weatherman."

—The JAG Hunter

Wherever one looks in this case, down through all these years of appeal, one notes a consistent unwillingness to probe for the simple truth.

20.
I AM MY FATHER'S SON

"Even in our sleep, pain that cannot forget falls drop by drop upon the heart until, in our own despair, against our will, comes wisdom through the awful grace of God."

—Aeschylus, *Agamemnon*
The first play of the
Oresteia Trilogy (458 BC)

"It is the same with any life. Imagine one selected day, struck out of it, and think how different, its course would have been. Pause you who read this, and think for a moment of the long chain of iron or gold, of thorns or flowers that would never have bound you, but for the formation of first link on one memorable day."[185]

—Charles Dickens, *Great Expectations* (1861)

⚓

THIS IS THE MOST SACRED action that I may carry out in my life: honoring my father! Obviously, this is a deeply emotional issue, and one that is very personal to me.

John Bitoff stole my name and at the same time, cruelly, he also stole my father's name.

So, forever intolerant, I am taking them back!

Two men whom I greatly admired in my youth were my father, Walter Francis Fitzpatrick Jr., and President John Fitzgerald Kennedy. I worshiped my dad. My sincere pride in my dad as well as my keen esteem for

---
185 Dickens, Charles. *Great Expectations*, p. 79.

President Kennedy were forged in my adolescence, accomplished with the clear knowledge that both men were navy veterans and combatants in World War II.

A graduate of Georgetown Medical School, Washington, DC, Walter Francis Fitzpatrick Jr. was commissioned a junior grade lieutenant of the Medical Corps in the United States Navy Reserve on 27 June 1941, six months before the attack on Pearl Harbor.

My dad was a combat doctor.[186]

At the beginning of World War II, my father reported aboard the *USS Cole* (DD-155) as a medical officer homeported in Norfolk, Virginia. He joined the ship whilst she was in Rhode Island. Dad served on the *USS Cole* (DD-155) from December 1941 to May 1943.

Ensign Edward C. Hines Jr. reported aboard in March. With hostilities escalating in Europe, graduation was moved up from June to February; he was a Naval Academy graduate, Class of 1941. Commander Hines gave me an oral history of the more than three years that he and my dad were shipmates.[187]

The *Cole* conducted convoy duty for many months with port visits in Boston, Iceland, Argentina, and Newfoundland. My dad was with me only for a short time. When he passed, he was only forty-nine, and I was only eleven.

I loved my dad. Even now, I miss him every day.

My memory of him is scant. I guard recollections of him that are strong and intense, but also quite sparse.

Sometimes I wonder if Dad had done more to raise me after he passed than if had he lived a longer life.

I joined the Boy Scouts, eventually achieving the rank of Eagle Scout, because, for whatever reason, I entertained the mistaken notion that my dad had been an Eagle Scout himself!

186 My dad was a combat doctor.
187 Commander Hines. January 1998. Oral history, East Carolina Manuscript Collection.

Dad did not talk much to myself or my younger brother. In truth, he was distant and sullen most of the time, and often he was quite insular. I did not understand then his embittered condition, and what since then I have come to recognize as a closed-in, beleaguered psychological condition.

Dad enjoyed watching *The Phil Silvers Show* (1955–1959) and particularly enjoyed the character of Army Sergeant Ernie Bilko. *The Soupy Sales Show* (1953–1966) was another favorite of his, and Dad never missed an episode of Hennesey (1959–1962) with Jackie Cooper playing Charles "Chick" Hennesey, a navy doctor based in San Diego. These television shows stand out because they always made my dad laugh. I do not remember Dad laughing any other times.

I remember when Dad broke down and cried uncontrollably one night after learning that his father, my paternal grandfather, had died. That day was the only time I saw my dad cry.

Dad was a dedicated Yankees fan. The only childhood remembrance I have of a family outing was the four of us going to see the Yankees take on the Angels at the old Wrigley Field (the one in Los Angeles, not in Chicago!) built in 1925 at Avalon Boulevard and 42nd Place, just southwest of Downtown Los Angeles. It must have been 1963, and John Kennedy was spending his first summer in the White House as president. That day, as was then the custom, my dad wore a formal suit, tie, and fedora to the game, and we all thoroughly enjoyed the event.

⚓

Left to me from my father's various naval belongings was a copy of *North Atlantic Patrol: The Log of a Seagoing Artist*. And folded in half, stuck behind the front cover, was a 1943 *Saturday Evening Post* article about America's invasion into Northwest Africa and Casablanca. It would be several years before it occurred to me that the "navy doctor" reported attending to a near mortally wounded army soldier on page 76 was actually my father: Walter Francis Fitzpatrick Jr.

Beforehand, I had known some of the command history of the *USS Cole* (DD-155), enough to discover the role the *Cole* had played during Operation Torch on 8 November 1942. The *Cole's* speedrun into Safi Harbor as part of the Western Task Force's incursion onto the Atlantic coast of Morocco, along with the USS Bernadou (DD-153), was described as a "suicide mission." Operation Torch was the first WWII incursion of US forces into the European-North African Theatre.

Years later, not finding anything more of my father's military record, I wrote to the Bureau of Naval Personnel (BUPERS) requesting copies of papers I thought existed. Unfortunately, I was informed that Dr. Fitzpatrick's military records were among the sixteen to eighteen million records destroyed in the 12 July 1973 St. Louis archives fire.

Now, decades later, while doing research for another book under construction, I wrote to Navy Curator and Historian Rear Admiral Samuel Cox in April 2017 with questions about the *USS Indianapolis* (CA-35). I chanced to advance to him a separate question: whether there might be any slim remnants of my father's records held by navy historians that might have survived.

Gladly, Rear Admiral Cox's team miraculously discovered Dad's military biography (essentially Dad's DD-214) and more—the initial write-up for the Bronze Star with the combat valor device "V," awarded for performance of duty under fire during the North African suicide amphibious assault.

Now, with gratitude and appreciation extended to RADM Cox's research team, I have learned that my dad served with the forces afloat aboard the *Cole* for eighteen months, from December 1941 to May 1943.

Before the *USS Cole* was reassigned and reconfigured for the suicide mission, the ship was a convoy escort in 1941–42 on the North Atlantic Patrol during the Battle of the Atlantic.

Dad sailed on the *Cole* for five convoys, escorting merchant ships to Newfoundland and Iceland from June 1941 to January 1942. From March through September 1942, the vessel patrolled the waters of the East Coast and did one escort run I know of to the Virgin Islands (convoy #6 before Operation Torch).

Dr. Fitzpatrick signed the back of the front cover of his copy of *North Atlantic Patrol* and then had his wardroom shipmates sign on the facing page. Lieutenant Commander George Palmer signed, as commanding officer of the *Cole*, and then Ensign E. C. Hines also applied his signature.

Rear Admiral Cox's find informs us that *USS Cole* shipmate Ensign Hines put my dad's name in for the Legion of Merit. Commanding Officer Palmer upgraded the nomination to the Bronze Star with "V."

Captain Palmer was awarded the Navy Cross for the same action. The *Cole* crew members were distinguished with an award of the Presidential Unit Citation (PUC).

After Operation Torch, the *Cole* sailed home to the States. There, the ship took up convoy escort assignments once more for a short stint of three months (December 1942–February 1943).

Eight of the convoys the *Cole* sailed, with Dr. Fitzpatrick assigned as convoy medical officer, were done so in waters thick with German submarines.

The *Cole* returned to the Mediterranean by way of Gibraltar in March 1943, and then, steaming farther into the Mediterranean, the *Cole* docked in Algeria in May 1943. I think it was in Mers El Kébir in Northwest Algeria that Dad finally transferred off the *USS Cole*.

Next, Dr. Fitzpatrick served in the Pacific Theatre during America's last WWII battle: the 1945 invasion of Okinawa, also known as Operation Iceberg.

Lately, I have learned of another possible indirect connection to Dr. Fitzpatrick: I have learned the special augmented hospital concept was developed in the late summer of 1944 in America's planning for what was to be the invasion of Japan's Home Islands. It appears to me that these special medical units were developed and trained in the same fashion as the construction battalions created, trained, and deployed earlier during WWII. The Navy Bureau of Medicine and Surgery (BUMED), if you will, is the answer to *The Fighting Seabees*, thereby creating doctor and corpsman one minute, combatants the next.

Enlisted personnel began training in San Bruno, California. I have come to understand that other plans, preparations, and training were also carried out in Newport, Rhode Island. Dad was assigned to Special Augmented Hospital #3 as assistant medical officer in October 1944 in Newport. It is not a long reach, given my father's extensive combat experiences in the North African battle actions, to consider that Dr. Fitzpatrick was integral to the startup of the "augmented hospitals."

Specialized Seabee battalions were a major step in building the special augmented hospital. (See Morrison, Samuel Elliott. (1945). *Victory in the Pacific*. 1945 (XIV). p. 276.)

Please consider the brutal humanity thrust upon servicemen that my father must have seen for all those long years as a navy doctor. Countless civilian Japanese would have gone through these hospitals. Surely, my dad took care of many of these patients as well.

Japan declared an "honorable surrender" on 2 July 1945 (see Morrison, p. 276, Barwise Report). Winston Churchill's verdict regarding the Battle for Okinawa, in a solemn message to President Truman on 22 June 1945, said: "The strength of willpower, devotion and technical resources applied by the United States to this task, joined with the death struggle of the enemy [...] places this battle among the most intense and famous in military history. [...] We make our salute to all your troops and their commanders engaged."[188]

Edward L. Beach observed: "By system of measurement,

---

188 Morrison, Samuel Eliot. *History of United States Naval Operations in World War II: Victory in the Pacific*. Vol. XIV. Little Brown and Company, Boston, 1960, p. 282. Morrison cites a *New York Times* article.

Okinawa was the bloodiest and [longest-lasting] battle in the history of any navy."

Dr. Fitzpatrick was transferred as executive officer to special augmented hospital #8 in October 1945. Dad returned Stateside in 1947. During the Korean conflict, he returned to the Western Pacific for a ten-month deployment with the Military Sealift Transportation Service (MSTS) from September 1950 to July 1951.

The attached formal portrait of my father was taken in 1947 at a studio in Newport, Rhode Island, just after Dr. Fitzpatrick had returned home from a twenty-six-month tour in Okinawa (1944–1947).

The last battle ended on Monday, 2 July 1945, when the US Navy, Marine, and Army forces secured the occupation of Okinawa. Operation Iceberg, the Okinawa campaign, was officially declared over on 2 July 1945.[189] However, the battle was not over for my father.

My most memorable day, for it made great changes in me, was when I

---

189 Morrison, Samuel Eliot. *History of United States Naval Operations in World War II: Victory in the Pacific.* Vol. XIV. Little Brown and Company, Boston, 1960, p. 276.

lost my father when he, by his own hand, brought his wartime nightmares and mental struggles to an end. That action took place precisely eighteen years later during the early morning hours of Tuesday, 2 July 1963.

There was no funeral mass for my dad. There were no monuments.

Understandably, my strong father's memorial, I did not fully grasp for many years. Perhaps he was suffering from what we would today term post-traumatic stress disorder (PTSD), and, with that realization in hand, perhaps he could have gotten counseling immediately after the war's conclusion—counseling which would have discouraged suicide and which might have encouraged him to live a normal, long life. There is no doubt that my father had wounds, deep wounds, from all the gruesome and horrific battles that he had witnessed, even if one could not see those wounds on the surface.

I think today of all the rough scenes he must have witnessed! My father was a combat doctor who made six convoy runs during the Battle of the Atlantic; he was awarded a Bronze Star with "V" for the *Cole's* suicide run into Safi, Morocco, during Operation Torch; he sailed in at least two more Battle of the Atlantic convoys afterwards (totaling a minimum of eight convoy runs); he was awarded two Presidential Unit Citations; he was part of two invasion forces (during Operations Torch and Iceberg); and finally, he was awarded four WWII campaign battle medals. Lastly, he took part in a WESTPAC deployment during the Korean conflict.

The most sacred thing I will do in my life is honor my father:

Medical Doctor and Navy Captain

Walter Francis Fitzpatrick Junior.

I am my father's son.

When signed formally, my name is

Walter Francis Fitzpatrick III.

This is my tribute to my father:

"May the road rise up to meet you.

May the wind be always at your back.

May the sun shine warm upon your face;

the rain fall soft upon your fields

and until we meet again,

may God hold you in the palm of his hand."

This is the most well-known Irish blessing, originally written in Gaelic, the original language of Ireland, and reliably attributed to Saint Patrick.

21.
THE ADMIRALS' GESTAPO!

"Never give in, never give in, never, never, never, never—in nothing, great or small, large or petty!"

—Winston Samuel Churchill
Harrow School, October 29, 1941

"The [military government] was never set up to ensure justice. [...] It is set up as your servant, a servant of the civilian population of this country to do a particular job [...] and that function [...] demands [...] almost a violation of the very concepts upon which our government is established."

—Five-Star General of the Army
Dwight David Eisenhower, 1948[190]

⚓

SIR MATTHEW HALE DECLARED IN 1645: "[Military] law, which is built upon no settled principles, but is entirely arbitrary in its decisions, is, as Sir Matthew Hale observes, in truth and reality no law, but something to be indulged rather than allowed as law. The necessity of order and discipline in an army is the only thing which can give it countenance."[191]

⚓

Congress enacted a military law on May 29, 1830, that "provided that when a military commander is authorized [...] to 'appoint' a general

---

190 Littell, Robert. (1970). "Military Justice on Trial." *Newsweek*, p. 18.
191 *Blackstone's Commentaries*, Vol. I, p. 413. Also see: William Winthrop, *Military Law and Precedents* (reprinted version, 2000), p. 47.

court-martial, the 'accuser or prosecutor' of an officer of his command proposed to be brought to trial, the court shall be appointed by the President.

"Its purpose clearly was to debar a superior from selecting the court for the prosecution and trial of a junior under his command and, as reviewing authority, passing upon the procedures of such trial, or executing the punishment, if any awarded him, in a case where, by reason of having preferred the charge or undertaken personally to pursue it, he might be biased against the accused, if indeed he had not already prejudged his case."[192]

And Colonel Winthrop informs us:

Preferring charges, in a general sense, consists in being the author of, or person responsible for, specific accusations presented against an officer or soldier [or sailor]. The "accuser" referred to in [the Articles of War and Articles for the Government of the Navy] is, in this sense, the preferer of the charges; and so is the "prosecutor" where he has either originated or adopted the accusation. In *law*, however, and as the term is employed to the present connection, the preferring of charges consists in the formal subscription and authentication of such charges for official purposes.[193]

Colonel Winthrop further points out the object of the 1830 provision was "to prevent the packing of a court, and still more perhaps to prevent the suspicion of such packing."[194] Additionally, "where it is held that if the convening commander was accuser or prosecutor, the court was 'illegally constituted, and the findings and sentence consequently void.'"[195]

He goes on to write, "An officer executes the sentence of a military

---

192 Winthrop, p. 62.
193 Winthrop, p. 153. Also see footnote 32.
194 Winthrop, p. 62. Also see footnote 33.
195 Winthrop, p. 62. Also see footnote 33.

tribunal which was without jurisdiction, or whose proceedings or judgment were otherwise illegal so that the sentence is invalidated, is a trespasser, and liable to an action for damages on the part of the person sentenced."[196]

In addition to a given military criminal act being judiciable under the UCMJ, such a cause may also be raised up in civilian federal court under the *Dynes v. Hoover* doctrine that, as we have seen, establishes federal judicial oversight of courts-martial on these two grounds: (1) it did not have jurisdiction over the subject matter or charge, and (2) it failed to observe the rules prescribed by the statue for its exercise.[197]

These prohibitions notwithstanding, it was the navy secretary's outlawed and peculiar practice in the nineteenth century to prefer charges against subordinate naval officers that in that day had "no counterpart in the military service."[198]

In more recent times, the practice of unlawful command influence (UCI) described above metastasized throughout defense department environs, and unsurprisingly, it is regularly exercised by commanders to "grind" up their subordinates. Absent due enforcement, the military law and federal criminal law against UCI atrophied since, in practice, a law not enforced is no law at all. Therefore, unlawful command influence has become a common and accepted military establishment practice. Recalling our title, clandestine Acts of Supremacy have become ordinary over time. The *"Functio imperium, lex est quod dico"* precedent prevails undisturbed: "Controlling function, the law is what I say it is."

⚓

---

196 Winthrop, p, 882. Also see Winthrop's footnote for cases cited.
197 Generous, William T. Jr. Editor James P. Shenton. *Swords and Scales: The Development of the Uniform Code of Military Justice.* Kennikat Press, 1973, p. 165. Also see: *Dynes v. Hoover*, 61 U.S. 65, 15L. Ed. 838, 20 How. 65, 1857 U.S. LEXIS 432. Also see: Kastenberg, Joshua E. *The Blackstone of Military Law: Colonel William Winthrop.* The Scarecrow Press, Inc., 2009, p. 137.
198 Winthrop, p. 153. Also see footnote 32.

Every small detail in this recitation of the facts of my case is already known to federal law enforcement officials, NCIS agents (formerly NIS when Bitoff first took off after me), and the most senior civilian and uniformed Navy Department and Defense Department officials. All the clear and obvious facts are known to Rear Admiral John Bitoff and his sycophants. They are also known to every navy TJAG since Schachte moving forward, and that includes key members of the weaponized FBI.

⚓

As the holders of some papers still being held in secret, some characters within my case obviously still know more than me.

⚓

Here is some "inside baseball" regarding the storied NCIS.

Gerald Nance and Leon Carroll Jr. each served as commissioned officers in the US Marine Corps. Both men left the marines to serve full carriers with the NIS/NCIS. For a period, the two men worked together as partners. Both individuals were later elevated to senior executive service (SES) positions.

In retirement, Carroll has worked as a technical advisor for over 370 episodes of the *NCIS* television franchise since 2003 and is credited as a screenwriter for one episode.[199] Show producers named one of the prominent characters in the show, Director Leon Vance, after the former NCIS special agent technical advisor Leon Carroll. Actor Rocky Carroll plays the part of Director Vance, but Rocky Carroll and Leon Carroll are not related.

Leon Carroll and Gerald Nance both played prominent roles in the NIS/NCIS court-martial cover-up crime drama central to this book.

---

199 Rudolph, Ileane. "The Real Criminal Minds." TV Guide, 8 December 2008, pp. 34–35. Also see https://www.imdb.com/title/tt0364845/.

⚓

From 1990 through 1993, while greatly distracted, I sought remedy and relief through navy JAG and various agency inspector generals. I was trying to get the court-martial conviction overturned, I was intent on continuing my navy career, and I also needed to be reassigned. However, matters only got more serious then, if not distinctly worse.

In arraigning for a new duty station, I experienced great difficulty. I had to force orders to be issued to the USS Carl Vinson (CVN-70) by relocating my family from the San Francisco Bay Area north to Bremerton, Washington.

When I reported aboard, Captain Doyle Borchers II promptly charged me with being in an unauthorized absence status, stood me before the Captain's Mast, confiscated a month's pay, and compelled the convening of an administrative separation board.

Borchers was a monster!

I understand today that Borchers had worked with Bitoff and other navy personnel to attack my character, to inflict further harm and injury upon me, and finally, to eject me from the navy altogether with an unfavorable discharge characterization.

The initial collection of documents found in this work were discovered while I was trying to forestall forced separation, thus jeopardizing my twenty-year full retirement with benefits.

That newly discovered evidence was used as the basis for a seventeen-page formal criminal complaint filed with the NCIS on 23 September 1993.[200]

⚓

---

200 Twenty-three page NCIS criminal complaint dated 23 September 1993.

I personally met with NCIS Special Agent Mark Sakarada on that date and delivered the formal criminal complaint in person in the NCIS office aboard Treasure Island, San Francisco. Separately, the report was widely distributed throughout the navy. Navy Secretary John Dalton signified his receipt on 14 October 1993.

Dalton further disseminated the criminal complaint to Navy Undersecretary Richard J. Danzig (USN), to S. Honigman, his office of general counsel (OGC), Navy TJAG Harold Eric "Rick" Grant (JAG), NCIS Director Roy D. Nedrow (NCIS), and a special assistant for legal and legislative matters (SAL), with specific orders to "HOLD CLOSE!"

The criminal complaint was subsequently guarded as a national secret until well after my forced retirement in September 1994.

⚓

In March 1997, I worked for a military contracting firm in Alexandria, Virginia, in the Crystal Gateway North building in Crystal City next to the Ronald Reagan International Airport and the 14th Street Bridge to the north. The Pentagon was just to the west, across Interstate 395.

The mid-Atlantic Defense Criminal Investigative Service (DCIS) field office was then located in the same building where I worked. My office was on the eighth floor, and the DCIS office was on deck below on the seventh floor.

By late March 1997, I had given up on the NCIS investigating my complaints. After all, I had been turned away so many times! So, I collected up my briefing materials and descended the stairway one deck.

I announced myself to the receptionist sitting behind a thick window. Special Agent Helena P. Wong was summoned, and I was allowed entrance to an interview room.

I was fifteen minutes into my brief, profiling the Anderson forgery, when Special Agent Wong was able to visually discern that the writing was counterfeit, at which time she called in another male agent.

I started over with my brief, repeating the same account to the male agent, who gave me neither his card nor his name. He did, however, upon visual survey recognize the bogus simulated application of my signature, thus proving that the signature was, indeed, a forgery.

Shocked, the two agents alerted their boss, Larry Leonard, who was the senior special agent in charge (SAC) of the DCIS mid-Atlantic office. He, in turn, joined the interview as well.

SAC Leonard, too, was shocked. He was increasingly so, especially when I clearly and insistently narrated the key role that TJAG Rick Grant and Grant's assistant, LCDR Nanette DeRenzi,[201] had played in later systematically blocking any criminal inquiry into Rear Admiral Bitoff's rigged court-martial.

However, despite realizing how radioactive my evidence was, SAC Leonard punted, gave up, deferred action, and abdicated any responsibility, as so many other well-paid officials had done before him. Leonard explained the NCIS was responsible for investigating. Leonard said my situation fell squarely within NCIS jurisdiction. Leonard told me he had a friend who was highly placed in the NCIS organization. As he then showed me the door, Leonard finally assured me that his NCIS point of contact friend would be in touch from his Washington Navy Yard office very soon.

The guy Leonard knew was NCIS Special Agent Gerald Nance, deputy assistant director for criminal investigations and one of the most senior NCIS agents in the organization, working in the third level of management under Deputy Director for Criminal Investigations Ernie Simon. Thomas A. Betro was the NCIS director at the time. All these

---

201 I didn't know at the time that "Nan" DeRenzi was a shooting star in the navy JAG community. In addition to her job with TJAG Grant as an appellate attorney, DeRenzi held the job of Navy Secretary Dalton's special assistant for legal matters. She promoted to vice admiral and became the first female navy TJAG. DeRenzi's success was no doubt attributable to her performance of duty assisting SECNAV Dalton and TJAG Grant to keep the lid on the Bitoff court-martial scandal. RADM Grant, in concert with SECNAV Dalton, blocked the NCIS from touching my case throughout Grant's and Dalton's entire tours of duty, Grant from early 1993 to early 1997 and SECNAV Dalton from. DeRenzi was a sycophant quisling for both.

positions are "SES" pay grades. The senior executive service is a civil service ranking equivalent to flag rank grades, or admiral, in the US Navy.

Larry called his buddy Gerald Nance, known as Gerry, and then Gerry called me a few days later on 3 April 1997. Before the call, Nance had taken into his possession and reviewed "five large three-ring binders so full that they can hardly close." Nance had also spoken to the new Navy TJAG John Hutson, who told Nance he "[would] reconsider [Fitzpatrick's] appeal if it can be established the document is a forgery."[202]

But Nance believed the signature was truly mine when he picked up the phone to call me. At first, Nance was sure I was crying wolf by reporting the forgery of my name—something done in a desperate attempt to draw NCIS attention to my voluminous other criminal accusations naming Bitoff and Zeller.

In that phone call, Nance was harsh in both tone and language. Nance zeroed in on the reported forgery confession and was apparently confirmed in his belief that I had signed the questioned document myself. He warned me that should the NCIS initiate an investigation, they would be coming after *me!*

Nance tried to talk me out of pressing my accusations, most especially my report of the forged confession. He explained that he and many other NCIS officials were simply tired of hearing about it. Simply put, they were fed up with me! As the phone call progressed, Nance became even more insistent, even threatening.

Nance stridently tried to compel me to buy into the odd and silly notion that I was lying, even saying that agents would find the "confession, then polygraph [me] to prove [I was] lying." Eventually, he hissed that they would find the writing of interest, prove it to be my genuine signature, have me recalled to active duty, and finally have me court-martialed again for a second time, this time tossing away the key to my cell on a second federal conviction.

---

202 Gerry Nance's 3 April 1997 internal memo.

Nance was real nasty. He wanted me to back off; however, his intimidation did not work.

I stood my ground. And then I told Nance: "Knock yourself out."

Then I briefed him. And somehow, mysteriously, just at that instant, a strange and unpredictable sea change ensued.

Not disclosing it to me, sometime during my report Nance reversed course. Nance *changed his mind!* He wrote later that day, "My opinion: [Fitzpatrick] is telling the truth, nothing like this could be a dream."

With TJAG Grant in retirement, it was on 3 April 1997—after being muzzled for a period of over four years since 14 October 1993—that the NCIS finally and first opened a criminal investigation into one feature of Bitoff's criminal expedition. NCIS Special Agent Richard Allen caught the case.

Assistant Deputy Director Nance's 3 April 1997 benchmark writing recording the initiation of NCIS efforts to recover the original court-martial record is instructive on these two key points:

- Nance held and reviewed five volumes of credible documentary evidence filed with various NCIS agents over the years.

- Nance gave Allen the go-ahead to contact a US attorney, yet he gave no specific guidance regarding how to proceed in the event that forgery was proven.

Special Agent Allen called my office at 10:35 hours local (ET) on 23 April 1997 to schedule an interview appointment and left a voicemail message. I returned his call the next day, and the interview was set for 30 April 1997 in my office.

NCIS Special Agent Richard Allen drove to my Arlington, VA, office and took my sworn statement on 30 April 1997.[203] As DCIS had acknowledged the month before, Special Agent Allen recognized by visual inspection that the document in question did not bear my true signature. Praise be to God: Allen, like Nance, believed me! What is more,

---

203 My statement dated 30 April 1997.

Allen found the massive volumes of documented evidence credible. Allen stated that the next step would be to examine the original court-martial record. A simple enough action—or so it was thought on 30 April 1997.

Then, I heard only silence.

Four months passed. May, June, July, August all came and went, and nothing was heard.

Allen told me nothing. There was no progress report.

Fed up with more unexplained delays, I decided to take action. It was 5 September 1997, a Friday morning, when I walked into Rear Admiral John Hutson's Pentagon office. Hutson wore two stars on his collar, holding the position as the 36th Navy Judge Advocate General.

I stepped into the small office and announced myself and my purpose to the second-class petty officer sitting at a small reception desk just inside to my left. A navy captain sat on a small couch, backed up against a bulkhead, a bit farther back to my right.

The navy captain jumped to his feet the instant he heard my name. He interrupted my exchange with the petty officer and blocked me from advancing any farther into the room.

Rear Admiral Hutson was in his office that day. Swiftly, the captain interdicted my attempt to meet with Hutson.

At the time, Don Guter was Navy Captain John Hutson's deputy (Navy TJAG under instruction). Guter knew me.

So, I did not meet with John Hutson on that Friday afternoon; Guter made sure of that. He backed me out of the office, one step at a time, into the expansive corridor, all the while lamenting that TJAG Hutson was not available. I remember him wearing a Band-Aid on his forehead over his left eye, and I remember beads of sweat trickling down over the covering. Guter did *not* want Hutson to become aware of my physical presence in the outer office.

Guter gave me a "wave off," nervously redirecting me to see Captain Rand Pixa, the Navy Judge Advocate General's Corps inspector general. Pixa was a "good friend," and someone who would be detailed to join with the NCIS in a companion effort to find the missing court-martial record. Pixa's primary duty was as head of the navy JAG's Admiralty Division. His assignment as navy JAG inspector general was a collateral duty.

In a desperate state after I left, Guter and Hutson immediately and urgently contacted Ernie Simon. Simon was then an NCIS assistant deputy director for criminal investigations, a senior executive service (SES) billet. Gerry Nance was Ernie Simon's deputy.

No one saw me coming on that Friday morning in September! Gerry Nance and Richard Allen had each determined I was telling the truth back in April 1997. Special Agent Allen had seen the document in question and just as the DCIS team had recognized the criminal instrument, the fake confession, as a forgery, so did Richard Allen. Then, after his interview with me, Allen searched for the court-martial record, which should have been immediately available, but the record had gone missing. Then, after another four months' delay, I unexpectedly came knocking at TJAG Hutson's door demanding answers.

Hutson, Guter, Nance, Simon, and Allen all knew the NCIS had been in a desperate and thus far unsuccessful search for the court-martial record without any explanation as to why it disappeared or any clue as to where it might be found. All concerned were fully cognizant of Nance's interaction with me as well as Allen's interview results, which had only strengthened the idea that my accusations were legitimate. Finally, the veracity of my criminal filings skyrocketed when no court-martial record could be found anywhere.

High-ranking civilian and military officials recoiled when they understood the motives behind the Secretary of the Navy Dalton's and TJAG Grant's aggressive and obsessive obstruction. Grant was gone; however, Dalton was still in office. The door-stopper volumes of credible, unarguable, and inculpatory evidence were in plain view right over there, stopping doors: five large binders' worth! I was telling the truth and asking the right questions, and people in very high places had suddenly become very frightened. The consequence of confirming that the questioned document was indeed a forgery was starting to take hold throughout the navy.

Fueled by both real and perceived dire outcomes, therefore, a real panic arose. Navy brass fully recognized the gravamen of the situation. The dire need for a cover-up and damage control now colored the tone and tenor of the criminal investigation that Gerry Nance had launched in April 1997.

Special Agent Simon knew what confirmation of the forgery meant. When alerted by the surprise TJAG Hutson and Deputy Guter call on a Friday afternoon, Simon expressed his appreciation for the seriousness of the situation and the consequent need for an immediate cover-up. Alarmed and triggered regarding the counterfeit "confession" bearing my forged (and misspelled) name, Simon sounded "general quarters," exhorting the other senior military governors to the very real danger that the forgery manifested for the image of the navy and the marine corps. In other, simpler words, all hands knew I was telling the truth.

Simon expressed his real panic regarding the dangerous character of the accepted forgery in an NCIS secret internal memo. In it, he telegraphed the necessity for a cover-up in his alert to all NCIS agents since he, Simon, was concerned that the forgery was an extremely sensitive piece of writing requiring special handling.

In 1981, then Secretary of the Navy John H. Leman once said, "[...] and the flag officers had their own police agency, the Naval Investigative Service (NIS), which Leman called the Admirals' Gestapo." Visita, Gregory L., 1995. *Fall From Glory: The Men Who Sank the U.S. Navy.* New York, Simon & Schuster, p. 24.

In truth, this is what Leman meant:

From: NCIS *DEPUTY DIRECTOR ERNIE SIMON*
To: 27HOST_DOM:25HOST_DOM:SRVHQOO_DOM:DCWAHOST_DOM: *SUPERIOR OF SA*
Date: Friday, September 5, 1997 12:17 pm   *RICHARD ALLEN*
subject: FITZPATRICK -Reply -Reply -Reply -Reply
            RICHARD
As far as I know, *ALLEN)* picked up the entire history when he
interviewed Fitzpatrick. Qucik and dirty, he was a Naval Officer who
was court martialed early 90s. Although retained in the Navy, he
ultimately was retired at a grade lower than he expected due to the
court martial. He blames a JAG officer for all his woes and claims
every flag offiers who reviewed his case and took no action is part of
a conspiracy, including *RADM H-R. "BILL" GRANT (OTJAG)*

Over the years, he has corresponded with OJAG, NCIS, Congress and
everyone else along the way to no avail. Many people in the front
office have had contact with him to incude I believe, *DOROTHY (DIRECTOR BRANCH SECD*
SA CHUCK BRANT   (?), *SA GERRY NAKE* and myself. *SA NAKE* finally
agreed to OPEN the forgery case you are working. Again, the forgery
is only one allegation of many Fitz has made against the Navy.
However, if you can prove the forgery, it totally supports his 10
years worth of contentions and makes the NAV look really bad. The
front office resurfaced this case to OJAG and they agreed to take
another look at it since *RADM GRN* has now departed. Fitzpatrick called
the front office last week to seek a meeting with the Director. He
was fended off but arrangements were made for *CAPT ALA* to talk to him
next Monday. *CAPT ALA* is trying to get the latest on what you're doing
so he's prepared for the discussion.

have some other e-mails I'll forward to you for background.   b7C

RECEIVED
TUESDAY, 11 AUGUST 1998
NCIS FOIA RESPONSE

FILED
OCT 1 4 2004
DISTRICT COURT
BY_____

Simon wrote:"[...] the forgery is only one allegation of many [LCDR Fitzpatrick] has made against the Navy. However, if you can prove the forgery, it totally supports [Fitzpatrick's] 10 years' worth of contentions and makes the [navy and marine corps], the NCIS, the FBI and [the] congressmen 'look really bad.'"

This also made every navy TJAG since Schachte, members of the FBI, and every congressman "look really bad" as well.

SPECIAL AGENT IN CHARGE NCIS PACIFIC **b7C** NORTHWEST

From: LEON CARROLL, JR.
To: W.\OFF1CE31\GRP\PSFO.GRP
Date: Monday, May 4, 1998 1:23 pm
Subject: OFF LIMITS

This is a notice to all Puget Sound personnel:

A Mr. Walter Fitzpatrick has made inquiries regarding an event
that happened to him several years ago when he was the XO of the
USS Mars homeported in San Francisco. He was administratively
discharged from the navy and is know claiming he was framed.
While residing in the Washington, DC area, he made a complaint to
our DC office that a memo with his signature forged was used in
the proceedings. DCWA opened a case and had the handwriting
examined and the results were inconclusive. There is nothing
more we can do for Mr. Fitzpatrick. He has now levied charges
against his former defense attorney, now a deputy prosecutor with
Kitsap County alleging that it was the attorney who forged his
signature. Again this has been investigated and the case closed
by DCWA. **b7C**

This morning Mr. Fitzpatrick arrived at the Bremerton Office to
file the same complaint. Fortuunately RAC MARY CALL was familiar
with the situation and explained to Mr. Fitzpatrick that NCIS had
looked into his complaint and could do nothing further to help
him. He departed NCISRA Bremerton stating he was going to take
his case to U.S. Rep. Norm Dicks.

Mr. Fitzpatrick has shopped his story around for years and has
reached the point of shere separation. He is not to be allowed
access to any of our spaces. If he shows up or calls your office
politely tell him that NCIS has looked into to his complaint and
any further information regarding his case can be obtained by
quering our headquarters through the Freedom of Information Act.

b7C

cc:

RECEIVED

TUESDAY, 11 AUGUST 1998
NCIS FOIA RESPONSE

It made the NCIS, the FBI, and the congressmen look bad too!

DEPARTMENT OF THE NAVY
NAVAL CRIMINAL INVESTIGATIVE SERVICE
PUGET SOUND FIELD OFFICE
9857 LEVIN ROAD N.W. SUITE L20
SILVERDALE, WA 98383-9406

Ser PS/102-98
05 August 1998

Mr. Walter Fitzpatrick, III
825 NE Rimrock Dr.
Bremerton, Wa 98311

RECEIVED
FRIDAY 7 AUGUST 1998

Dear Mr. Fitzpatrick:

This letter is in followup to our meeting the late afternoon of
15JUL98. As stated to you at that time, the decision to further
investigate your allegations would be made by Naval Criminal
Investigative Service (NCIS) headquarters in Washington, DC. All
information provided by you to me was forwarded for
consideration. I have been told that you have been informed by
NCIS headquarters that no further investigative activity will be
pursued based on the information provided to date. As promised
during our 15JUL meeting, I am returning to you all documentation
provided that day and all subsequent faxes/documents. Any
subsequent inquiries should be addressed to NCIS Headquarters
without information copies sent to this office. Should NCIS
headquarters decide on any further investigation regarding this
matter, the appropriate field element will be tasked.

Respectfully,

L. Carroll, Jr
Special Agent in Charge

Three days after Ernie Simon's urgent missive, Special Agent Richard Allen filed a status report subject: "ATTEMPTS TO LOCATE ORIGINAL DOCUMENT."[204]

⚓

Guter did order Navy JAG Inspector General Pixa to act; however, that was a head fake—just another misleading ruse. Pixa called me a few days after 5 September. I traveled to Pixa in Hoffman Building II on a handful of occasions, always to be deterred and disappointed in the lack

---

204 Special Agent Allen's status report dated 8 September 1997

of progress. I was intentionally kept in the dark all along, not once briefed on the status of the search for the missing court-martial record.

September 1997 came and went. During the four months between 30 April and 5 September 1997, initial attempts to find the original court-martial record failed. Bitoff and Zeller had filed the counterfeit instrument in the record, but the record was not where it should have been. The record was last seen when Marine Corps Major JAG Richard K. "Rick" Stutzel checked the volume out of the Federal Records Center (FRC) in Suitland, Maryland, on 7 December 1992. Major Stutzel had returned with the court-martial record to his office situated aboard the Washington Navy Yard, just four blocks away from the home to the chief of naval operations (CNO).

Then, the record vanished from sight.

It is either funny or ironic that in 1993, Stutzel worked in the very same building housing the main headquarters of the NCIS.

⚓

During the five years of its concealment, between 28 April 1989 and 12 August 1994, Chief of Naval Operations (CNO) Mike Boorda personally reviewed the court-martial record.[205]

Boorda's involvement was spurred by Ed Offley's *Seattle Post-Intelligencer* article, published two days before Boorda was sworn in as CNO, and a companion editorial published four days after Boorda took the helm. At the same time, Congress was finally applying extreme pressure, asking and compelling Boorda to deliver some decent answers.[206] In response, Boorda wrote in his own hand: "I reviewed the case very carefully." Boorda said he "wanted to make sure [he] had all the facts before responding."

In that simple statement, Boorda simply lied in his response to

205 Boorda's letter to Congressman Dicks.
206 Ed Offley's *Seattle PI* article with companion editorial.

Congress. He was at once obedient to Secretary of the Navy Dalton's 14 October 1993 gag order,[207] and further, he was working to conceal his own treacherous mendacity.

Containing the original copy of the forgery—a forgery reported as a crime to the NCIS on 23 September 1993—Boorda maintained the court-martial record in secret. He kept it under tight control and seal until he took his own life on 16 May 1996, aged fifty-six, in the backyard of his Washington Navy Yard home.

After Boorda's death, the court-martial record was returned to Major Stutzel and then buried once more.

⚓

Special Agent Allen's search expanded but secured no result. Allen then filed a status report respecting efforts to locate the original documents on 8 September 1997, three days after my visit to TJAG Hutson's office and three days after Ernie Simon's panicked internal memo.

I first met with Navy TJAG Inspector General Captain Rand Pixa the following Tuesday, 9 September 1997, in his Hoffman Building II office in south Alexandria, Virginia. I met with Pixa many times only to be given the runaround every time, and sadly Richard Allen had also gone underground.

May, June, July, August, September, and October passed into November 1997, and I heard nothing by way of progress from either Allen or Guter. By this time, I knew for sure that once again, I was being strung along.

Undeterred, and maybe somewhat bull-headed and stubborn, I yet again turned to the FBI for an independent investigation. In the three-page criminal complaint filed with FBI Special Agent Weaver in November of 1997, I accused NCIS officials of "criminal facilitation by

---

207 SECNAV Dalton's cover sheet.

failing to investigate my charges when first formally presented [on 23 September 1993] and since."[208]

And yet again, predictively, there was absolutely no FBI response.

⚓

I left the Washington, DC, area in November 1997, returning to Bremerton, Washington.

However, exactly one month later, Pixa discovered the court-martial record had been "misfiled" at the Navy-Marine Corps Appellate Review Activity (NAMARA) and NCIS main headquarters aboard the Washington Navy Yard on Thursday, 4 December 1997.[209]

Special Agent Allen called me at my Bremerton home on 5 December 1997. He verbally reported that Captain Pixa had disinterred the original court-martial record from a cardboard box found in a closet in an office formerly occupied by USMC Major Richard Stutzel. Allen went on to relate that the original copy of the forgery had been seized as evidence, along with several other matching documents. These additional matching records were printed using the same paper stock, font size, type, and style and were all documents Kevin Anderson had signed using his own name.[210]

Together, Pixa and Allen surveyed the seized original counterfeit confession the next day on 5 December. Allen seized a number of additional documents as evidence in the ongoing criminal investigation as well.

With the counterfeit in hand, Allen was able to detect similarities between the paper with my simulated signature and other papers in the same record. Upon close inspection, the forgery felt the same as other papers that stood apart as different from other entries, as the text on these similar writings was identical in typeface, type font, and type. Holding

---

208 Complaint filed 3 November 1997.

209 "FITZPATRICK UPDATE" dated 5 December 1997

210 NCIS memo dated 5 December 1997.

these matching documents up to the light, one covering the other, some words appeared in both while others in the underlying copy disappeared, completely obscured by the topmost rendering.

The matching writings all bore Kevin Martis "Andy" Anderson's true signature—the one that appeared on the forged document where Anderson had signed my own name.

In that instant, Anderson replaced Tim Zeller as the lead suspect in the crime under federal investigation: the forgery of my name. Anderson, a captain of the marines JAG, was the officer Admiral Bitoff and Lieutenant Zeller had detailed to me as my assigned defense counsel. Anderson had been my defense attorney for the court-martial.

WOW! Imagine my shock!

NCIS criminal investigators had slammed headfirst into the sophisticated, elaborate, and secreted Bitoff/Zeller court-martial criminal enterprise.

So, on 5 December 1997, my long-standing accusations were not only totally verified, vindicating me completely, but now the criminal investigation that Nance had launched back in April—after initially embracing the mistaken theory that I had signed the "confession" myself—had boomeranged and burgeoned to the point of becoming overwhelming. Bitoff's handpicked defense counsel, Kevin Anderson, had emerged as the lead suspect.

Due to the recovery of the original papers, NCIS investigators were then forced to interview Kevin Anderson. Horrified, Captain Pixa called from Washington, DC, to warn Anderson in Washington State that the NCIS would soon be knocking on his door—and Pixa also sent Anderson a copy of the newly unearthed forgery.[211]

⚓

---

211 "Results of Interview" dated 26 January 1997.

Rand Pixa disinterred the original court-martial record from its cloaked hiding place in a cardboard box in a locked office closet at the Washington Navy Yard, just steps away from the official residence of the chief of naval operations. The closet was one that Major Rick Stutzel had used during his time in the building where the Navy-Marine Corps Appellate Review Activity was situated on the first floor and the NCIS was headquartered on the second.

⚓

Quickly, NCIS Special Agent Allen turned the forgery and matching documents over to the NCIS Regional Forensic Laboratory in Norfolk, Virginia, on 8 December 1997.[212] NCIS investigators obtained document examiner Fred Dudink's handwriting laboratory report, dated 22 October 1997, and that report made a comparison between my true handwriting and that of Zeller's. Mr. Dudink declared the memorandum to be a forgery, but he cleared Zeller of the crime.

No effort was ever made to directly obtain witnessed samples of Kevin Anderson's handwriting. Papers signed by Anderson using his own name—available in the court-martial records that Special Agent Allen had seized as evidence (read: the matching documents)—were ignored.

The NCIS Norfolk forensics lab filed its report on 19 December 1997, reporting only the comparison of my true handwriting to Zeller's.

It was determined that a fabric felt-tipped writing instrument had been used. Anderson, for one, regularly used a black Skilcraft pen (pictured).[213] No fingerprint analysis was attempted.

NCIS investigators also intentionally neglected to apply other industry standard investigative techniques to pursue in plain view or to consider obvious clues. And please take special notice of this key fact: NCIS

---

212 "Recovery of Original Document" dated 8 December 1997 and "Case Summary" dated 30 December 1997.
213 Fountain pen and Skilcraft pen.

investigators did not—I say again did *not*—examine papers incriminating Kevin Anderson, such as the matching documents Anderson had signed using his own name. Incredibly, Zeller's handwriting was scrutinized and compared to the forgery, but not Anderson's.[214]

Again, NCIS agents did not examine or analyze the original documents that were identical matches in all aspects to the original forgery, all of which Special Agent Allen had seized from the court-martial record. Senior NCIS criminal investigators intentionally eliminated Kevin Anderson from scrutiny as a suspect. Thus, the fix was in.

In the meantime, Kevin Anderson had left the marine corps. He now worked as a Kitsap County assistant prosecuting attorney in Port Orchard, Washington. Port Orchard is the county seat, abutting Bremerton and across the Puget Sound from Seattle. Port Orchard is located inside the Pacific Northwest (PACNORWEST) NCIS regional footprint. Leon Carroll Jr. was the special agent in charge in 1997, officed in nearby Silverdale, Washington.

The NCIS interviewed Anderson in his Port Orchard office on 26 January 1997, three weeks after the NCIS lab report was completed. Anderson was not questioned as a suspect, but rather as a possible source of information for the origination and administration of the forgery.

During the interview, Anderson said he "was certain" that information stated in paragraph four of the bogus writing had been passed to me, though he did not explain how I had been so informed. Anderson further intimated, as Zeller had, that I had signed the criminal instrument myself. Beyond that, Anderson said "he did not recall the memorandum [read: forgery]" and claimed he did not apply my simulated signature to the writing.[215]

Because Anderson was not under suspicion, the NCIS interviewer(s) failed to ask Anderson for his further cooperation by submitting

---

214 NCIS lab analysis report dated 3 February 1998.
215 "Results of Interview" from Kevin Anderson's interview dated 26 January 1998.

handwriting examples himself. Anderson did not volunteer to submit to a handwriting examination as Zeller had.

The day after the NCIS Anderson interview, in an exchange between unidentified (names redacted) senior NCIS leadership, it was written: "[...] our charter (SECNAV NST 5520.3B) specifically says that NCIS can defer investigations '(w)hen in NCIS judgment, the inquiry would be fruitless and unproductive.' I'd say this would qualify."[216] Therefore, NCIS governors shut their investigation down completely on 3 February 1998.[217]

And that is precisely when I came out of my skin! NCIS agents had Anderson dead to rights with the stark discovery of the matching documents on 5 December 1997—and then, unaccountably, they worked assiduously thereafter to quietly walk away from the issue.

PACNORWEST Regional SAC Leon Carroll Jr. knew that Anderson had forged my name. Thus, Leon Carroll feared that the scandal was about to erupt.

Alarm bells rang everywhere.

On the morning of 4 May 1998, I went to the NCIS office aboard the Puget Sound Naval Shipyard in Bremerton, Washington to try to inspire a reopening of their investigation of Kevin Anderson. I was only minutes into my interaction with Special Agent James H. Connolly when Connolly's boss, Resident Agent in Charge (RAC) Mary Call, barged in and ordered me out of their spaces.

Special Agent Call then alerted Leon Carroll in his Silverdale office.

Carroll then sent out notice to all Puget Sound NCIS personnel that their offices were "OFF LIMITS" to me.[218] Separately, Secretary of the Navy Dalton had recommended to the US Senate Tim Zeller's promotion

216 "WALT FITZPATRICK" internal memo dated 27 January 1997.
217 NCIS Report of Investigation (Closed) dated 3 February 1998 by NCIS Deputy Assistant Director for Congressional Affairs P. Cole Hanner. Letter to Senator Slade Gorton (R-WA) dated 30 March 1998.
218 Leon Carroll "OFF LIMITS" memorandum dated 4 May 1998.

to full commander, stating that the NCIS had cleared Zeller of all alleged misconduct.[219]

Congressman Dicks was continuously updated that the NCIS had abruptly shut down their investigation, the object of which was Anderson's forgery. Dicks asked the Secretary of the Navy Dalton to take the matter up personally in his letter dated 16 July 1999.

Still, in all of these twists and turns, I refused to give up! I pressed hard for a one-and-one meeting with Carroll and he strong-armed me until 15 July 1998, whereupon Carroll abruptly told me "to go pound sand." Carroll returned all proof and evidence that I had left with him under a cover of a letter dated 5 August 1998.[220]

I was all over Katrina C. Pflaumer (US attorney for the Western District of Washington), Representatives Norm Dicks, Senator Slade Gorton, and local FBI agents. Pflaumer, completely derelict in her duties, washed her hands of the entire matter in mid-October 1998.[221]

Meanwhile, busy crickets chirped away from all points of the compass.

⚓

With the information I had available to me in the mid-nineties, it was my belief at that time that Tim Zeller had forged my name. As it happened, Zeller was as much a part of the creation and management of the bogus writing as Kevin Anderson was; it was just that Zeller did not actually put pen to paper.

With an eye toward going around the navy's stonewalling, I contacted Zeller's Oklahoma Bar Association (OBA). My complaint lay dormant for years until one of the OBA investigators, Tony Blasier, thankfully took a studied interest. Mr. Blasier was a trained and licensed investigator and document examiner, and he also was someone qualified to administer lie

219 Dalton's letter to assistant SECDEF dated 11 June 1998.
220 Leon Carroll's letter dated 5 August 1998.
221 US attorney Pflaumer's letter dated 15 October 1998.

detector tests.

It was while Tim Zeller was under Investigator Blasier's intense scrutiny that Zeller commissioned Mr. Fred Dudink to conduct a handwriting examination of Zeller and myself. Dudink's report was released in October 1987 and taken from Zeller as evidence by NCIS investigators.

I informed Mr. Blasier immediately just as Kevin Anderson had been revealed as the person who had forged my name. In 1998 and into 1999, I petitioned the NCIS to make the original of the forgery available for Mr. Blasier's and Fred Dudink's independent examination.

That very normal request was denied.[222]

It was not until 2001 that a certified true copy of the counterfeit was turned over, and that action was a direct result of Tony Blasier's dogged pursuits. I remain most thankful for those actions.

⚓

Still, I did not relent!

In October 1998, I reached out to Congressman Norm Dicks again. Navy Secretary John Dalton had again reported the NCIS criminal participation, complicity, and cover-up upon discovery of Kevin Anderson's outlaw expedition with Bitoff and Zeller. However, Dalton waited to toss the "hot potato" to his successor, Richard Danzig, who then tossed it to the Assistant Secretary Carolyn H. Becraft, who eventually punted the ball to Navy TJAG John Hutson. At this point, exasperated beyond description, I asked myself a simple question: Where is a leader who owns some spine?

Eventually, Hutson responded, saying: "[...] repeated investigations have failed to demonstrate any substance to your many and long-standing claims of improper treatment. For these reasons, your requests cannot be granted."[223]

222 C. C. Briant's letter dated 28 October 1998.
223 TJAG Hutson's letter dated 20 November 1998.

Next, I pushed hard on Admiral Bitoff himself to come forward and admit what he had done—to the point where Bitoff actually threatened me that if I did not stop, he, Bitoff, would file a report with the FBI against *me*.

Bitoff backed off upon review of just a few documents sent to him.

⚓

Throughout the years, Congressman Dicks was continuously updated. Dicks waited over a year to inquire about the NCIS decision to abruptly shut down their investigation, the object of which was Anderson's forgery. It took Admiral Bitoff's 30 April 1999 letter to Dicks to force Dicks to act. Admiral Bitoff's letter called out Anderson's forgery with laser beam focus.[224]

Dicks asked Navy Secretary Danzig to take the matter up personally in a letter dated 16 July 1999. Part of Danzig's answer, explaining why no further action was appropriate, needs to be read and considered:

Norm—

I am very much influenced by Mike Boorda's 1994 review and that of various other reviews.

Richard[225]

Unbeknownst to Dicks, Danzig had Bitoff's letter to Dicks in hand since Bitoff had sent Danzig a copy a month earlier on 4 June 1999.[226] Bitoff sent his letter to Navy TJAG John Hutson.

Danzig was briefed three times in June and July 1999 about the forgery of my name and the NCIS criminal facilitation in the cover-up. In his response to "Norm," Danzig made no mention of anything regarding John Bitoff's earlier outreach calling out Kevin Anderson's criminal adventure.

---

224 Appendix one: Bitoff's letter to Norm Dicks
225 Danzig's 20 September 1999 letter to Norm Dicks
226 Bitoff's letter to SECNAV Richard Danzig.

Danzig's answer to Bitoff, if there was one, was secret.

⚓

At this point, I sought civilian legal representation. Immediately, I was mocked, laughed at, and advised that the Feres Doctrine represented an absolute block to any civil or criminal lawsuit. The Feres Doctrine prevents members of the armed forces who are injured while on active duty from successfully suing the federal government under the Federal Tort Claims Act (FTCA).

Then, for a number of years, I contacted numerous civilian defense attorneys whose clients Kevin Anderson was prosecuting. I told these various attorneys that Anderson had forged my name and was, therefore, a criminal at large.

In the fall of 2004, I was in the Kitsap County Courthouse Sheriff's Office reception area when a sheriff's sergeant overheard one of my conversations.

Congressman Dicks wrote to FBI Director Louis Freeh on 28 March 2001, requesting the FBI open a criminal investigation into the forgery.[227] On March 33, I paid an unannounced visit to NCIS headquarters aboard the Washington Navy Yard in DC. I identified myself to the attendant guard and asked to meet with NCIS Director David L Brant. The guard called for the "suit" security officer, NCIS Special Agent Ron Bell, who arrived on the scene to promptly eject me from the building.

The next day, after both giving testimony to the Commission on the Uniform Code of Military Justice, Glenda Ewing and I paid a surprise visit to Norm Dicks's Washington, DC, office.[228] I told Representative Dicks about the maltreatment that Special Agent Bell had given me the day before in my effort to meet with Director Brant.

Days later, Dicks petitioned the FBI for an independent investiga-

227 Norm Dicks's letter dated 28 March 2001.
228 "Citizens Against Military Injustice" (CAMI) article.

tion—outside of the military establishment—into the outlawry of Admiral John Bitoff, Bitoff's pirate associates and protectors, and the NCIS criminal participation and facilitation.

Then, one more time, crickets were heard—only crickets!

⚓

In June 2001, I rode the ferry across the Puget Sound to Seattle to report Kevin Anderson to the Defense Criminal Investigative Service (DCIS). Special Agents Henry G. Mungle and Christina M. Mihaltse took my sworn, verbal, and written petition. Jennifer Wallace was the resident agent in charge.

In my introduction, I explained that this was my second attempt with the DCIS; further, I recounted my earlier March 1997 visit with Special Agents Wong, Leonard, and mystery agent #3.

Immediately, Mungle and Mihaltse accused me of lying to them and threatened federal prosecution targeting me.[229] Then seven more months went by, and again, nothing happened!

In late January 2002, Resident Agent in Charge Jennifer Wallace walked away.

⚓

I recovered the police report on Friday, 17 September 2004, reading it while still at the police station in Port Orchard, and then I drove directly to Kinko's in nearby Silverdale to make copies—*many* copies. Next, I traveled to the nearby NCIS office and presented myself at the window in the small reception area. No one was at the window. From this vantage point, a person is able to look into a much larger office area. No one was to be seen. There was no bell or device to use to make it known there was a visitor standing at the reception window.

---

229 DCIS engagement of 25 June 2001: Mungle/Mihaltse.

Still reeling about my moment's old discovery, I took copies of the police report and taped them to the window facing inward, covering then entire window as far up as I could reach. Then I used more copies to wallpaper the reception area. Then I left.

L. G. Beyer, NCIS inspector general, responded to my aggressively unorthodox delivery of Deatherage's report in a letter dated 30 November 2004.

On the afternoon of Thursday, 3 March 2005, NCIS Special Agent James H. Connolly led a team of four (Connolly plus three) to my residence in Bremerton, Washington. The NCIS team positioned themselves in a defensive perimeter, holding the front door as the focal point in a potential crossfire of their killing field.[230] Then Connolly, unannounced, knocked on the door.

I got up from a sick bed to open the door.

Connolly and his NCIS gang were there to deliver a warning and a letter.[231] NCIS Inspector General L. G. Beyer had signed out the letter, dated 30 November 2004, which contained a number of proven lies. Connolly warned me that I was risking my life should I continue to try to expose Bitoff and his criminal cohorts in their criminal expedition.

Connolly's and Beyer's meaning was clear: *Do not go there, Fitzpatrick. Stop. Do not proceed. Otherwise, we are going to kill you.* Connolly held the letter and did all the talking. Connolly was about six feet four and stood to my right at the door. A second shorter agent stood tensed up to my left. Two more agents stood in the middle distance, arms hanging to their sides, hands crossed at their wrists in front of them, coiled in readiness to react. Each of the four agents were armed.

By this time, top NCIS management was in a panic because Anderson's police report had exposed everything—not only Bitoff but everybody. And still, there was more to discover that I did not then realize.

---

230 "Connolly Deployment."
231 "Naval Criminal Investigative Organization" dated 30 November 2004.

⚓

LIEUTENANT CULPEPPER'S DEFENSE

"The [court-martial] went off all right with all the precision of a well drilled cast doing a well-rehearsed play, the [court-martial] looked fine, up to the very last minute. The three witnesses told their stories clearly and simply, as if quoting their typescript statements from memory; their stories all jibed. The prosecutor explained with incontestable lucidity the infractions of the [Articles of War] that had been committed and the penalty required by the [Articles] that had been committed and the penalty for such infractions. [...] Everything looked rosy, everything was according to Hoyle. Then, at the last moment, with a sort of abortive outrage against destiny, Lt. Culpepper suddenly entered a furious plea of guilty and appeal for clemency on the grounds that all good soldiers were drunkards [...].

The accused could gladly have shot him."

—James Jones, *From Here to Eternity* (1951)

⚓

US Attorney John McKay was aggressively working on implementing the new Law Enforcement Information Exchange Program (LInX) in 2004, an advanced information-sharing system sponsored by the navy and various law enforcement agencies across the country.[232] United States

---

232 Iglesias, David with Davin Seay. *In Justice: Inside the Scandal That Rocked the Bush Administration.* John Wiley & Sons, Inc., 2008, p. 108. Also see: Statement of John McKay, Former United States Attorney For the Western District of Washington, Before the *Subcommittee on Intelligence, Information Sharing and Terrorism Risk Assessment*, Committee on Homeland Security United States House of Representatives, September 24, 2008: https://fas.org/irp/congress/2008_hr/092408mckay.pdf.

Deputy Attorney General James B. Comey detailed McKay to oversee the pilot program.

However, Justice Department infighting had defied McKay's funding requests; the LInX project was floundering. The NCIS special agent in charge (SAC) headquartered in Silverdale, Washington, Scott Jacobs, came to the rescue to try to right the ship. Using Navy Department resources, Jacobs provided LInX with the funding that the Department of Justice had been reluctant to deliver. The NCIS took the lead on the information exchange prototype within the Pacific Northwest NCIS regional footprint, home to a nuclear-powered submarine base as the initial test bed. SAC Jacobs was the program manager, working closely with US Attorney McKay.

Kitsap County, Washington, lays entirely within the NCIS LInX envelope. Kevin Martis "Andy" Anderson was a Kitsap County assistant district attorney under Russell Hauge when Anderson uttered his sworn statement to Port Orchard Police. District Attorney Hauge sat as a witness in the Anderson/Detective Deatherage interview, and Anderson still held the position when his police interview was unearthed.

Clearly, much was at stake.

Every dirty detail revealed in these pages, and doubtless more, stood to be publicly exposed.

Neither the NCIS SAC Jacobs, US Attorney McKay, nor Kitsap County District Attorney Hauge could afford to address the perceived and looming scandal for their respective agencies—a scandal that criminal-at-large Kevin Anderson's revelatory admission promised.

Certainly, the trustworthiness and integrity of the NCIS were greatly at stake just eight years after the scandal-ridden NIS had changed its name to the NCIS in a reorganization spawned by the Tailhook atrocity that took place in September 1991.[233] The CBS *NCIS* television franchise

---

233 Vistica, Gregory L. *Fall From Glory: The Men Who Sank The U.S. Navy.* Touchstone, 1997, p. 420, fn 20.

launched in 2003. The show was very popular, enjoying great ratings, and its audience was growing.

The four NCIS agents who assaulted me worked out of the same Silverdale, Washington, office as SAC Jacobs.

NCIS criminal complicity in covering up Bitoff's expansive criminal conduct was not going to be allowed to jeopardize the burgeoning success of the LInX program.[234] Therefore, they kept the lid on.

On 17 January 2006, navy officials awarded the highest civilian award for his "innovative leadership" in developing LInX.[235] SAC Jacobs received the same recognition and was promptly promoted.

⚓

I filed my criminal complaint against Connolly with the local Silverdale office of the FBI (with Special Agents Patrick Gann and Stephanie Gleason) since this ran the risk of focusing bright sunlight on the operation of the NCIS in local communities (local policing) and on their overall operation. More than that, my criminal complaint opened the door on an investigation into a much larger criminal business, and it was sure to seriously threaten, if not quash outright, the planned launch of the LInX program. So, predictably, the FBI (DOJ), the NCIS, and the Department of Defense Inspector General (DODIG) all ignored and stonewalled the complaint of a felony crime.

⚓

The language found in the closing two paragraphs of NCIS Inspector General L. G. Beyer's letter is frenzied. Beyer was telling me, "DO NOT GO THERE!" Clearly, I had a document that was very dangerous to

---

234 Link: https://www.ncis.navy.mil/Mission/Partnership-Initiatives/LInX-D-Dex/.
235 Rondeau, Sharon. "Military Corruption: Cover-Up, Death Threats, and Power Grabs," The Post & Email, 10 July 2013, https://www.thepostemail.com/2013/07/10/military-corruption-cover-up-death-threats-and-power-grabs/.

them, and Beyer was saying, "If you come back with that document, if you pursue this, we will find a way to arrest you and lock you up [the specter of Jerry Nance arisen]. We know what it means; we know what it says."

Yet, for Beyer, it was not enough to simply write the warning. So, he sent Connolly and his goon squad to the door of my residence to threaten me with my life. And all the while, I was wondering: *Is this the best that we can do? Is this the America I know and love?*

Still sick and bedridden, I cobbled together a written report the next morning accusing Connolly and his thugs of assault. I took it to the local Silverdale, Washington, FBI office, knocked on the locked door (during business hours), and waited.

No answer. I slid the complaint under the door and waited some more.

Nothing happened.

Then I took the written complaint to the NCIS office Connolly had warned me hours before never to visit. I asked to meet with someone from their "internal affairs" division, and I was sent away instead.

I kept going back. NCIS officials kept sending me away.

The other dimension to this moment was LInX.

THE FORGERY!

It is unnatural and ironic that you now possess a copy of the forged misspelling of my name, but sure enough, a copy is available in this book. That counterfeit confession was not ever supposed to see the light of day, since it was supposed to remain a secret forever.

And this is precisely how a court-martial is meant to work!

Rear Admiral John Bitoff, as the convening authority in my court-martial *and* accuser, partnered with his Staff JAG Zeller, but he needed a third accomplice to safely proceed. So, Bitoff ordered or assigned (then) Marine Corps Captain Kevin Martis "Andy" Anderson as my defense counsel.

This threesome, working in concert, secreted documents, destroyed documents, and manufactured, out of the clear blue sky, at least one document.

As a function of command, John Bitoff commissioned the creation of my "confession" [*sic*], something needed at the end of the court-martial process to cover his tracks and dissuade any serious after-action review. By the way, once again, there is no such thing as an "appeal" in the military discipline process—just senior command review.

Bitoff ordered Anderson to create the false "confession," forge my name to the confession, and then he instructed Anderson to turn it over back to Bitoff (via Zeller) for entry into the official record.

Marine Captain Kevin Anderson is the man who put pen to paper in the forgery. Anderson admits both in writing and under oath that he, Kevin Anderson, is the person who created the bogus writing. Signature samplers of my true signature are available in this book.

The reader will find that Kevin Anderson's signature samplers are also supplied. Carefully consider that the name "Fitzpatrick" ends with a *K*, not a with a *T*, as Anderson misspelled the forgery. Anderson also failed to add the generational suffix, the Roman numeral III.

⚓

Only Kevin Anderson touched the criminal writing from the beginning. I never saw Captain Anderson after my last meeting with him on the subject of the court-martial after 5 April 1990.

Incidentally, there is one more pesky detail that has not yet been revealed. There was never a question but that the attempted simulation of my signature was a forgery. I have never been in the same room with the original document of the bogus criminal writing.

The court-martial proper ran from 1–5 April 1990 in the Naval Service Legal Office aboard Treasure Island. When the hearing finished, I signed some paperwork for Anderson and then promptly departed. I did not see Anderson again for years. Further, I had nothing to do with any of Bitoff's staff or with Bitoff.

I was a stay-at-home dad from April through September 1990. I was checking in daily by phone with the Military Sealift Command duty officer where I had been temporarily detailed. My wife was pregnant. A scheduled delivery was set for 6 July 1990 at the Oak Knoll Naval Hospital. I did not know at the time that Tim Zeller's wife, Maureen, happened to be a navy nurse working at Oak Knoll.

In my absence, and completely without my knowledge or participation, Zeller presented Kevin Anderson with Bitoff's letter of reprimand (LOR) under a cover letter dated 11 June 1990. Zeller's communication announced a deadline for my response of fifteen days after my receipt.

Anderson held the LOR for twenty-seven days and recorded that he delivered the LOR to me on 7 July 1990.[236]

Anderson's lie is recorded.

I picked up my wife and new baby girl from the hospital and drove them home on 7 July.[237] Anderson did not deliver the letter to me, and I knew nothing about the fifteen-day deadline for response.

Zeller's nurse wife told her husband, who then told Anderson, the date of my wife's scheduled delivery on 6 July 1990. Surely, both Anderson and Zeller knew that I was unavailable to receive any documents on 7 July 1990.

Anderson waited ten days, until 17 July, to deliver Anderson's forged response to Zeller and Bitoff.

Again, I have never been in the same room as the counterfeit instrument. I did not even learn of the forgery's existence until May 1992 by

236 Forwarding and delivery cover sheet
237 Cathy and Angel

way of a response to a Freedom of Information Act request for the court-martial record.

Finally, over the years I have painstakingly conveyed all of this detailed information to every criminal investigator to whom I have ever reported.

⚓

On 5 September 1997, an NCIS internal memo recognized that the forgery of my name proves everything that I have reported—yet, beyond that, it proves so very much more.

There remained a great deal to cover-up.

17 July 1990

From: Lieutenant Commander Walter Francis Fitzpatrick, USN
To: Commander, Combat Logistics Group One

Subj: **RESPONSE TO LETTER OF REPRIMAND**

Ref: (a) Ltr of Reprimand, 7 June 1990
       (b) Record of Trial ICO US v. LCDR Fitzpatrick

1. In response to the letter of reprimand, reference (a), I would like to point out the following facts. It was the testimony of a government witness, Chief Wagner, the detailed MWR officer on board the USS Mars, that it was his suggestion to use MWR funds for the trip to Hawaii and that he advised me that this was an authorized expenditure. See page 53 of reference (b). The Master Chief of the command, PNMC Poasa Fa 'Aita, also testified that he saw a message about an MWR seminar to be held in Pearl Harbor and that he discussed this seminar with Chief Wagner, page 21 and 25 of reference (b).

2. When I used MWR as an excuse to fund a trip of ship's personnel to Hawaii it was because I was informed by my MWR officer that such a trip was authorized. If the convening authority believes this not to be the case then the convening authority chooses not to believe the testimony of the very witnesses called by the prosecution to testify against me.

3. In paragraph three of reference (a) I am reprimanded for the purchase of electronic equipment that was placed in my stateroom and the stateroom of the commanding officer. I would note that the record of trial, page 30 of reference (b), indicates that the commanding officer, Captain Nordeen, authorized the purchase and placement of this equipment. I would also note that Defense Exhibit "D" of reference (b) also establishes that the majority of ships in LOGGRU One currently place MWR entertainment equipment in officer's staterooms. It seems incongruous that I be reprimanded for an action taken by the commanding officer and which is conformance of the tolerated policy of the entire group. Furthermore, contrary to your statement in paragraph (3) of reference (a), I am not aware of any mention

252

in reference (b) to my having been "warned" against the distribution of entertainment equipment as was directed by the commanding officer. I would also note that reference (b) demonstrates that a small portion of the total funds was used for equipment placed in my stateroom and that of the commanding officer. Reference (b) also shows that this equipment replaced MWR equipment that was in place even before I arrived on board the ship.

4. I am also informed by my defense counsel that in his discussions with members of the court-martial panel it was disclosed that they did not consider to me guilty of dereliction in reference to the entertainment equipment. I believe this information is also known to your staff judge advocate.

5. I believe that I served the interests of the United States Navy and of the USS Mars well and to the best of my ability. I rely upon testimony of my commanding officer and the command master chief in evaluating my performance on board the USS Mars. Finally I rely upon the performance of the USS Mars while I served as her executive officer. I believe that my judgment, performance and dedication to duty did contribute to the service of the finest ship afloat in LOGGRU I.

Walter Fitzpatrick
LCDR, USN

I am confident that the kind reader now fully appreciates what the forgery proves, and also why, even today, key Department of Defense officials want me to be either silent or dead. The original of the forgery lurks somewhere in an office in Washington, DC. Last I heard (some years ago now), it was locked up in a safe in navy JAG offices aboard the Washington Navy Yard.

Theft, no matter how small, is prosecutable. Forgery—also known as identity theft—committed by any person is a crime. That is, except in the case that the crime is committed by the commander in chief (known to all of us as our chief executive, the president of the United States) or any of his very senior military governors.

I shall leave it to the individual reader to consider the larger dark consequences resulting from the operation of one government within another, both in competition with each other. Also, I repeat here and extend my exposition of the commander in chief as himself a criminal in command if the pursuit of true justice is not followed by his rigorous example.

Indeed, I believe that this difficult and tangled story of my court-martial at the hands of powerful forces within the navy, a navy that I loved, demonstrates that for me, justice was thwarted at every step. Finally, I hold that my tale shows that, on the other hand, the navy has (in the present tense) actual, real contempt for the average sailor, or what the French might term *"le mepris."* Throughout this ordeal—one which includes both the court-martial itself and my many appeals to various authorities for help, as this chapter demonstrates—rarely did I meet one courageous man, one inclined to pursue action and not just talk, someone willing "to stick his neck out" in order to do the right thing: free an innocent man. That situation of weak resolve and eternal vacillation reminds me of these lines from Shakespeare:

"Cowards die many times before their deaths. The valiant never taste of death but once."

William Shakespeare. *Julius Caesar* (1599). Act II, Scene 2, Lines 32–33. (Caesar speaking to his wife, Calpurnia.)

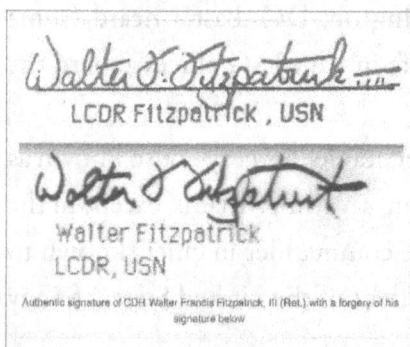

Authentic signature of CDR Walter Francis Fitzpatrick, III (Ret.) with a forgery of his signature below

The signature here is my true signature dated 5 April 1990. Anderson's attempted simulation of my name is below, dated 7 July 1990. It remains: "[...] because you can prove the forgery, it totally supports [Fitzpatrick's 30 years] worth of contentions and makes the [navy and marine corps] look really bad."

Then there is this:

thedrifter

Marine
Free Member
Join Date: Jun 2002
Location: Jacksonville, NC
Posts: 69,964
Credits: 21,693
Savings: 0

Former JAG attorney implicates JAG Corps

Former JAG attorney implicates JAG Corps
By Kit Jarrell
Dec. 31, 2006

Last night on the radio, Tim Harrington and I were talking about the JAG Corps, and the incredible web of lies and misconduct that permeates the inner workings of a system designed to maintain good order and discipline.

While documents I've posted speak for themselves, my conversation yesterday with former Marine and JAG attorney Timothy Zeller really brings it together. But first, some background.

Zeller, you may remember, was the orchestrator of a fraudulent court-martial against LCDR Walter Francis Fitzpatrick III in 1990. As the Staff Judge Advocate under Admiral John Bitoff, Commanding Officer of Combat Logistics Group One, Zeller's job was to act as Bitoff' chief legal adviser.

The problem is that Zeller was also one of Fitzpatrick's accusers, having signed the charge sheet. As the accuser, or "bringer of charges" against Fitzpatrick, Zeller had an obvious conflict of interest, since Bitoff had named himself the convening authority in the case. Bitoff, who ordered Zeller to sign the charge sheet on Bitoff's behalf, was also an accuser. Bitoff admitted his role as Fitzpatrick's accuser in a letter to Congressmen Norm Dicks in April 1999. Bitoff wrote to Dicks, "I brought the charges and I convened the court-martial" against LCDR Fitzpatrick.

Under both ethical and legal parameters, Bitoff wasn't allowed to assemble the Fitzpatrick court-martial but did anyway. Zeller and Bitoff were legally barred from any connection to the military hearing due to the fact that they were making the accusations against Fitzpatrick. As you'll see from the documents, Zeller gave all kinds of advice to Bitoff while the admiral conducted the proceedings, and it was exactly the kind of advice you'd expect from someone who desperately wanted to see Fitzpatrick punished by imprisonment. Bitoff and Zeller were working together.

After visiting the USS MARS--the ship Fitzpatrick was the XO of--in an investigative capacity, Zeller wrote a report directly to Rear Admiral John Bitoff. Dated 23 Oct 1989, the report recounted witness statements, made accusations, and discussed evidence. However, there were problems with this memo. A lot of problems.

Zeller fabricated testimony, claiming that parties were interviewed and gave damaging testimony against Fitzpatrick, yet in the list of evidence and statements in the report, none of these names appear, meaning Zeller never talked to them at all. Statements Zeller claims to have taken, and any information Zeller collected are completely missing. Zeller didn't see the need to maintain a chain of custody (standard investigative practice), inasmuch as Zeller didn't believe in keeping files to cover Bitoff and Zeller when their decisions might later be questioned (Zeller's policy statement to Bitoff in their "personal for" memo exchange of 11 April 1990).

Zeller claimed that financial documents for the MARS were missing, and but all found after the "departure of Operational Specialist Chief Wagoner." Zeller neglects to mention that Wagoner left the ship a year prior. The documents had never been missing at all. In fact, Tim Zeller was the last person to have the USS MARS 1988 fiscal year financial report, which contained exculpatory evidence that could have been used to exonerate LCDR Fitzpatrick. The the original was sent to Bitoff's CLG-1 command. Zeller seized remaining copies of the report into evidence. No version of the USS MARS' MWR financial report have been seen again.

Zeller makes the unequivocal assertion that "LCDR Fitzpatrick is guilty of dereliction of duty by failing to adhere to proper procedures for the expenditure of MWR funds," even though there had been no Article 32, no court-martial, and no verdict. This was simply one man, serving as the accuser in the case, reporting to the convening authority in the case, who just happened to be his boss.

Zeller's involvement in the case didn't stop there. He continued to advise Bitoff throughout the Article 32, the court-martial, and the subsequent disciplinary actions.

On January 9th, 1990, the Article 32 Report was signed by LCDR J. J. Quigley, the investigating officer. In Line 18, the report offers a "Yes" or "No" option to the statement, "Reasonable grounds exist to believe that the accused committed the offense(s) alleged." Next to the statement, Quigley had checked "No." However, two days later on 11 January, Zeller sent a memo to Bitoff advising him in a legal capacity that he could choose to reject the IO report and forge ahead with a general court-martial. In fact, the ever-helpful Zeller told Bitoff that "at the present time the Article 34 advice is being prepared in the event you desire to convene a general court martial."

During the trial, Zeller complained to Bitoff about the government's prosecutor, Matthew Bogoshian, claiming that the young attorney lacked the "desire to win." In 1992, Bogoshian gave a sworn statement that "the majority of charges LT Zeller brought against LCDR Fitzpatrick seemed to have little or no basis in reality." The prosecutor also mentioned that "LT Zeller seemed obsessed with the prosecution of LCDR Fitzpatrick."

After the court martial was over, Zeller once again advised Bitoff on the disposition of the case in a personal and confidential memo that was never made part of the public record. In fact, Zeller specifically assured Bitoff that there were no other copies of the memo. Zeller had no idea these memos would survive and be made public so many years later.

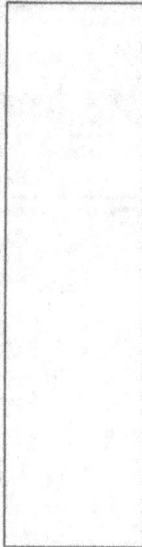

21. THE ADMIRALS' GESTAPO

There is more to this story...much more, including forgery and complicity on the part of the government defense attorney. In the next installment I'll continue the story. But allow me now to jump ahead to the present day, sixteen years later. Yesterday I spoke with Tim Zeller by phone.

Initially, he was friendly, professional. I gave my name and explained that the reason for my call was that his name had come up in the course of a story I was investigating concerning a case he had dealt with. He asked what case, and when I said Fitpatrick's name, Zeller laughed.

"That old thing has been going on for 16 years now," he said. "I'm not even in the Navy anymore. It keeps getting brought up, but nothing ever comes of it. If you want to go ahead and dig through all of it again be my guest. I don't have time for it. Have a good day."

"Before you hang up on me," I said, "I do need to tell you that I hold in my hand a number of memos that you wrote to an Admiral John Bitoff that are not part of the official record of the case."

The line went silent. "What memos?"

"Well," I went on, "there's this one, a report that you sent to Admiral Bitoff, stating that Fitzpatrick was guilty before the article 32 was even held. I also have one where you state that you don't like to keep copies of memos in case your actions are questioned later..."

He cut me off then, and his attitude changed significantly. "I've had psychiatrists tell me that I have to watch my back with that guy. I've had NCIS tell me the same thing. This is ridiculous..."

He hung up on me after reading me the Riot Act, but as I went back to writing, the phone rang again.

"Yeah, this is Tim Zeller. What's your radio show?" I gave him BlogTalkRadio's website, as well as Euphoric Reality, and offered him the number to call in to the radio show. He didn't want it.

"Did you talk to anybody at JAG?" he demanded. I told him that I've talked to a lot of people, that this was the culmination of months of research. I explained that I wasn't in the business of ruining people's lives or careers, and that this information had dropped into my lap during an investigation into the misconduct behind the Pendleton 8, Haditha, and Airborne cases. I again offered him the chance to come on the show and talk about the accusations, but he refused.

"I'm giving you a chance to have equal time," I told him.

"I haven't had equal time since this started," he sputtered. I told him I didn't know what to tell him, since I hadn't been involved "since it started."

"You even check your facts? You need to check your facts," he kept saying.

"Why do you think I'm calling you?" I asked. "I'm giving you a chance to answer this, to come on the show or call in and tell your side. There's evidence that this case was mishandled, and it's part of a bigger picture of misconduct on the part of the JAG Corps."

"I'll tell you, if you think this case was mishandled, I could tell some stories...I've been a defense attorney too." He paused, then, letting the unspoken hang.

"Well, I'll be more than happy to look at anything you might have on other cases," I told him.

"That depends," he said. "If I don't like your attitude when I go to your website, I'm not going to help you."

The conversation went downhill from there, and he ended up hanging up on me soon after.

The bottom line here is that there have been crimes committed by high-ranking officers in the military. These crimes have continued to be covered up throughout the years, and some of the same people who have perpetrated these acts continue to be in power today. They continue to handle cases, deal with potential witnesses and evidence, and decide the fate of sometimes innocent people.

Regardless of the political clout of these parties, their long and illustrious careers in the military, or their lucrative civilian employment today, their acts need to be exposed. It is not over dramatizing to say that the lives of men depend on it.

Quote

CREATE POST   Quick Navigation   HEADLINE NEWS   TOP

« Previous Thread | Next Thread »

THREAD INFORMATION

There are currently 1 users browsing this thread. (0 members and 1 guests)

POSTING PERMISSIONS

You may not Create Posts	BB code is On
You may not post replies	Smilies are On
You may not post attachments	[IMG] code is On
	HTML code is On
You may not edit your posts	Forum Rules

NEWS
Marine Corps News
DOD News
VA News
US News
World News
Politics

DISCUSSIONS
Leatherneck Locator
Open Squadbay
Slop Shute
Post (ee) Hall
Marine Corps
Recruits
Marine Corps Family

INTERACTIVE
Arcade
Chat Room
Locator
Marine Photos
Video
Social Groups

ABOUT
Advertising
About Us
Contact Us
Terms Of Service
Privacy Policy
Mac's Reactions

FOLLOW
Facebook   Twitter
               Youtube

CONTACT US MARINE CORPS - USMC COMMUNITY ARCHIVE PRIVACY STATEMENT ABOUT US TOP

Copyright 2001-2014 Leatherneck.com. All Rights Reserved.

http://www.leatherneck.com/forums/showthread.php?39351-Former-JAG-attorney-implicates-JAG-Corps   Page 4 of 5

The implications of this post stand for themselves.[238]

---

238 Thecreswellchronicle.com/news/story.cfm?story_no=4490.

EPILOGUE

"Nemo me impulse lacessit!"

"No one provokes me with impunity!"

— The Latin motto of the Royal Stuart dynasty of Scotland

⚓

"So, in the Libyan fable is it told,
that once an eagle, stricken with a dart,
said when the saw the fashion of the shaft,
with our own feathers, not by others' hand,
are we now smitter."

— Aeschylus (525–456 BC)
Fable 276 in the Perry Index

⚓

"Do you not know that you are the temple of God and that the spirit of God dwells in you? If anyone defiles the temple of God, God will destroy him. For the temple of God is holy, which temple you are. Let no one deceive himself. If anyone among you seems to be wise in this age, let him become a fool that he may become wise. For the wisdom of this world is foolishness with God. For it is written, 'He catches the wise in their own craftiness'; And again, 'The Lord knows the thoughts of the wise, that they are futile.' Therefore, let no one boast in men. For all things are yours: Whether Paul or Apollos or Cephas, or the world or life or death, or things present or things to come—all are yours. And you are Christ's, and Christ is God's."

— Saint Paul's First Letter to the Corinthians 3:16–23

⚓

"Nin sibi, sed patriae. Fiat voluntas tua."
"Not for self, but for country. For Thy will be done."

⚓

"We all know that there are rules and then there are rules. It all depends on whose ox is being gored, right? Or, one may ask: Who is enforcing the rules? Who has his hands on the reins of power? However, the real question is this one: Do we own the sheer will to find and secure true justice; or instead, shall we just look the other way and then always let whoever is the most powerful hold sway?"

CLOSURES

Any epilogue worth its stuffing must necessarily carry with it a two-pronged appeal for both greater courage and greater action, accomplished at the same time; otherwise, it is simply not worth neither the writing nor the reading.

⚓

The 17 November Group (17N) launched as a terrorist organization in 1975.

Criminals in command have been terrorizing America's service members and our US Constitution since 1775.

Gladly, we caught the renegade monsters from 17N. Yet today, command racketeers still roam about as freely as feral animals combing the untrammeled backwoods.

"If we should bump into one another, recognize me."

Let a freshly charged grand jury crew this ship.

Get your fire ship underway and proceed to sea!

Grand jury crew, set the special sea and anchor detail!

Your orders are to set sail, make first contact with, and engage the enemy.

Destroy the enemy by fire.

Use this ship as the torch that ignites a wild blaze, burning our unconstitutional government down.

Ensure your target is sufficiently afire.

Abandon ship.

Then let us rise up again from the ashes, phoenix-like, as a free nation once again.

⚓

USS MARS LAST MISSION

The *USS MARS* was commissioned as a Christmas present to the US Navy on 1 December 1963. The vessel was named for a community in Pennsylvania that bears the same name as the planet. The ship was the first in the class of new combat stores and named for the entire class of six vessels.

The *USS MARS* was transferred to the navy's civilian fleet one day before Groundhog Day on 1 February 1993 after serving in the active fleet for nearly thirty years. On that date, the *USS MARS*'s Military Sealift Command's distinctive blue and gold bands were painted onto her smokestack.

The *MARS* was finally decommissioned from all service five years later on 19 February 1998. Next, the ship's bell was removed before the *MARS* was towed out to sea from her Pearl Harbor mooring, and she was later used as a target during fleet exercises in 2006.

Finally, the *MARS* was sunk sixty-two miles (fifty-four nautical miles) off the coast of Hawaii in 2,750 fathoms (16,500 feet) of the Pacific Ocean on 15 July 2006.

The ship's bell proudly pays eternal tribute as a public memorial in the small city of Mars, Pennsylvania.

⚓

Walter Francis Fitzpatrick III and Robert F. Clifford were both proud members of Troop 508 as Boy Scouts in their youth in the sixties. With considerable focus and determination, both young men achieved the high rank of Eagle Scout.

Later in life, Walter and Robert both joined the navy, both serving as surface warfare officers in forces afloat. Robert was on active duty in 1988 when 17N murdered William Edward Nordeen. Bob soon left the service in 1989 to begin a career with the Federal Bureau of Investigation. 17N ran amok for many years after their murder of William Nordeen.

FBI Supervisory Special Robert Clifford, working as a legal attaché, assumed a leadership position with the joint terrorist task force in Athens, Greece, in November 2000. He personally led the investigation that resulted in the capture, prosecution, and successful conviction of more than a dozen dastardly leaders of the 17N terrorist cell.

⚓

As we round this, my run's last lap before the finish line, it is my firm and unshakeable belief that both uniformed and civilian Navy Department leaders have a solemn duty and responsibility to overturn the results of my manifestly unfair court-martial, to lift this biased federal conviction off my back, and to remove forever that horrendous stain on my naval reputation. There are a series of actions that must be taken, even if no formal prosecutions are pursued. After all this time, it is imperative that the record be set straight, and further, that the wayward treatment I suffered shall not be inflicted upon any other sailor or soldier in the future.

⚓

I am fully and eagerly prepared to testify in full painstaking detail before a federal grand jury regarding these complicated and pesky matters. I do hereby, by this writing, volunteer for that job, and a great deal more!

Many senior navy officials, both uniformed and civilian, know of Anderson's forgery, which remains an outstanding, unprosecuted crime. NCIS agents know of it. Those men and women fully understand that Rear Admiral John Bitoff rigged my court-martial from stem to stern. These same officials have a duty and a responsibility to cleanse my service record and to provide as much additional remedy and relief as they are capable. "Big Navy" must deal with this issue, which it entirely brought upon itself, no matter how lousy it makes the navy look to the general public.

I shall never forget any minute or trivial detail of this fiasco, one which so easily could have been entirely avoided. And no part of me, not even for a mere minute or tick of the clock, thinks that this despicable affair is over.

At this late juncture, perhaps it would be wise to recap the tawdry events and summarize *le deluge* one final time:

First, Rear Admiral John Bitoff received permission for an investigation from a three-star vice admiral named David M. Bennett (Commander, Naval Surface Force, Pacific).

After that, all bets were off.

Next, Bitoff took charge.

Then Bitoff illegally acted as both my senior investigation officer and prosecutor, acting as convening authority and saying, as we have seen: "I brought the charges, and I convened the court-martial."

Following that action, he dismissed the Naval Investigative Service (as it was known before it changed its name to Naval Criminal Investigative Service) from any inquiry. In doing so, he personally seizing an advantage, thereby cutting off any outside—presumably impartial—investigation.

Bitoff then pronounced me guilty of a crime (see Zeller investigation report).

Next, he assigned his own attorney to investigate and chose a lieutenant, forsaking the simple and universally accepted notion that a junior officer must not investigate a senior: Timothy W. Zeller. See Lieutenant Zeller's ominous Thanksgiving Day message saying: "We picked the defense counsel, Lieutenant Anderson, so that the hearing would be fair."

Next, John Bitoff ordered his assigned defense attorney to my case: Marine Lieutenant Kevin Anderson.

Bitoff then rigged the Article 32 investigation and hearing by threatening a former subordinate.

Then Admiral Bitoff rigged the court-martial, inserting his own staff, one who had been previously removed due to his acerbic prior conflicts with myself: Lieutenant Commander Steve Letchworth.

After the court-martial, Bitoff used a phantom witness, Lieutenant Commander Doug Dolan, to testify against me. Dolan was not interviewed by Zeller during his investigation report, a pertinent fact withheld from the court-martial. Dolan did not appear at my Article 32 hearing, either, and did not testify before my court-martial.

Lieutenant Anderson later forged my name to a confession. As we have seen, this is the same attorney whom Bitoff had assigned as defense counsel.

Lastly, John Bitoff tried to extract himself form all alienability, writing the fictitious, rear-end-covering letter above as soon as he realized that he might be close to being caught in extremis, that letter written to

Representee Dicks and Navy Secretary Danzig.

Duplicity, deviousness, and chicanery were practiced throughout! Instead, as readers and observers of this painful case, we must begin to act as if we were covered by a full and heightened system of justice—not by a runaway court-martial process that was, in my case, only a punitive tool for petulance, vengeance, and simple revenge.

⚓

In a final rearward view, let us reflect again upon the similar court-martial case of Captain McVay, the skipper of the *USS Indianapolis*. By this late juncture, I do pray that the reader has detected the many striking similarities between that very onerous case and my own. Once again, in like fashion, there is both a prosecutor and accuser in Secretary of the Navy Forrestal (1892–1949), illegally acting as the convening authority.

After the sinking of the *Indianapolis*, the navy was in trouble. Those in charge sought to deter and deflect criticism after its loss of a cruiser at enemy hands at the end of WWII, those desperate sailors having spent five days in shark-infested waters, with a total crew size of 1,197. 881 men survived the sinking, yet tragically, only 316 lived to tell the tale. Obviously, something had to be done! An example had to be made, and a convenient scapegoat had to be found—someone who could be blamed!

Thus, with tremendous and calculating cruelty, Forrestal took Captain McVay down, as he was a survivor, using the immoral tool of a court-martial.

Of course, as history tells us, Secretary of the Navy Forrestal was the true coward! The point is that the court-martial, for far too many instances through the centuries, has been used as a punitive or punishing mechanism rather than an instrument to find and secure simple, transparent justice.

In reality, as we look back into the past, the better to understand it, we come to understand that neither court-martial—neither mine nor Captain McVay's—consisted of anything close to court-martial material.

And in truth, the trying of both cases was an infraction of what today is known as the Racketeer Influenced Criminal and Corrupt Organizations Act.

Neither case was handled by *Dynes v. Hoover* (under US adjudication).

In both trials, there was no judicial oversight!

In both trials, revenge was paramount.

In both cases, there was no later judicial review.

Then, in both cases, a massive and complicated cover-up was then required, engendered, and then floated free for eager consumption by a lapdog press.

And through all of this, I maintain that these nefarious actions could only have taken place within a Godless society. As a court-martial runs its course, if the prosecution is unfair, some rules are adhered to and some are assiduously ignored, the better to suit that impartial prosecution and depending on the whim of the moment. However, in the hearts of all men involved in a court-martial trial, there must be automatically a contest between duplicity and revenge and, on the other hand, the simple desire for full and true justice.

Yet, we have seen what mean distractions and rampant evils take place when the military is not governed by the clear rules of God. We have seen what takes place in the military when a studied neglect of God becomes standard: vindictiveness and retribution may then soon reign. Still, after all that I have lived through in the thirty-plus years since my court-martial commenced, from time to time in the still of the night I do gather some greater strength by recalling to myself the Latin phrase : "Deus il vult." *God wills it.*

And at the very same time I remember these short, pungent words from John Dryden, which do begin this work: "Beware the fury of a patient man!"

Both short phrases give me the strength daily to go on until tomorrow and all the days beyond.

⚓

Daily, even now, I also draw even greater forbearance and tenacity from these words crafted by a great past president, one who, incidentally, in 1904 commissioned Naval Station Great Lakes, located just north of Chicago along chilly Lake Michigan in Great Lakes, Illinois:

> It is not the critic who counts [...] not the man who points out how the strong man stumbles, or where the doer of deeds could have done them better. [...] The credit belongs to the man who is actually in the arena [...] who spends himself in a worthy cause [...] who at best knows in the end the triumph of high achievement, and who at the worst, who, if he fails, at least fails while daring greatly, so that his place shale never be greatly be with those with those cold and timid souls who neither know victory nor defeat.

> —Theodore Roosevelt's "Citizenship in a Republic" Speech (also known as "The Man in the Arena" speech) delivered at the Sorbonne, Paris, 23 April 1910

⚓

At the close, it is well past high time that I present the Sailor's Creed in full:

> I am a United States Sailor. I will support and defend the Constitution of the United States of America and I will obey

the orders of those over me. I represent the fighting spirit of the navy and those who have gone before me to defend freedom and democracy around the world. I proudly serve my country's navy combat team with Honor, Courage, and Commitment. I am committed to excellence and the fair treatment of all.

That last line catches one's attention in that I do not believe (and the history of the telling proves as much) that I was treated fairly. Most of the time the current system of military justice works well, especially if there is little chance of judicial malfeasance, grim retribution, or simple monkeyshines. However, we must ask ourselves: What happens when the system fails? When a commander becomes unaccountably angry and vituperative, for whatever reason? What simple evil can take place when an innocent man is caught up, as I assuredly was, in untoward circumstances and is then sent packing to the meat grinder, his careers in tatters and his reputation irredeemably sullied?

Even to this day at this late writing, and though I have scratched my head on this issue countless times, I have no clear idea what small irritation may have sparked Admiral Bitoff's fiery animus towards me. I have no idea what trivial event might have catalyzed his persistent hostility toward me. For some unfathomable reason, I annoyed the admiral and he subsequently vowed to get me, to make an example of me, no matter what. Though I do scratch my head, I have no sure knowledge of the actual source of that incipient anger. Obviously, at some point I got on his bad side and, sure as anything, there I stayed!

Yet all the while, throughout this long and twisting tale, I loved the navy. I was most proud to be a member of it, and throughout my time in the navy, I tried my level best to do my job to the very best of my ability.

The theme of this book is that our military justice system desperately needs legislative reform and, in particular, to rework how a given court-martial is run. It should no longer exist under the exclusive purview of the defendant's commanding officer; the convening authority and the prosecutor cannot, in good faith, be the same officer. Any change to the rules rests within the legislative branch, and not the military itself. Article II, Section 2 of the Constitution states that an appeal may go straight to the commander in chief.

And finally, a permanent and fully compensated fifty-member board of oversight and review must be put in place to review any court-martial it selects. Such a board should be composed both of military officers and nonmilitary men and women of fairness—in other words, navy judge advocate generals and non-lawyers—and most importantly, it should have the clear and certain power to overturn the conviction of any soldier or sailor in a court-martial whenever it holds that simple fairness was not the overriding principle at work. That sort of perennial oversight, I do believe, would go a long way to ensure that future unsuspecting sailors or soldiers do not have to endure the kind of treachery and deception I have lived through.

However, unless circumstances change by the establishment of such a board of oversight and review, this dismal situation—one begging for unfairness to be wrought—reminds me strikingly of these words of another sailor, proud navy man President John Kennedy: "The rights of every man are diminished when the rights of one man are diminished." These prescient words were delivered in the middle of our nation's civil rights fractiousness on 11 June 1963.

I am reminded of another speech of his, too, one delivered to the citizens of West Virginia just before the election on 19 September 1960: "I think we can do better. I think we can do better. I think we are a great country, but we can be a greater country."

Lastly, I recall these striking words from prosecutor Jim Garrison in the movie *JFK*, directed in 1991 by Oliver Stone: "You know, going back to when we were children, I think most of us in this courtroom thought justice came automatically, that virtue was its own reward, that good triumphs over evil. But as we get older, we know that just isn't true. Individual human beings have to create justice, and that is not easy because the truth often poses a threat to power and one often has to fight power at great risk to themselves." (Screenplay from 1991, written by Oliver Stone and Zachary Sklar.)

"Individual human beings have to create justice"—why, that simple sentence could be the very theme of this book! Throughout these many years, in my endless appeals of my court-martial and the inevitable turn-downs when I was often threatened and neglected, commonly impugned and disregarded, as the reader may appreciate, I have had to fight the kind of stern power to which Prosecutor Garrison refers at great risk, both to myself and to my family.

In the end, I am grateful that God gave to me the constant and lastly strength to be able to have written this book. I am thankful that I have written it, too, an account which chronicles one man's long battle against an innately flawed military justice system. I believe that it proves to any-one with an unjaundiced mind that our military justice system needs a drastic and thorough improvement, including much greater judicial oversight and review, so that unwarranted mistakes like those visited upon me never take place again.

I am firm in my belief that right-minded officers will eventually see the validity of my claim—that the charges against me were fraudulent, orchestrated, trumped up, and then camouflaged—and further, that the rules regarding any court-martial will be fundamentally rewritten and improved.

If not now, when? Surely, that improvement would not be a thing easily done. If we wanted to improve the system, we could do it, if only we would first garner the sheer will to get that simple job done. One must ask the question: Can we not put our collective heads together to create a better mousetrap—one which would be fair and impartial in all instances, and to all soldiers and sailors—if that is what is meant to be?

I do know one thing for certain: If there is a true desire for justice, to go towards the light rather than toward the darkness, it shall be found. For as Saint John writes to all of us who do truly search for the light:

> Just as Moses lifted up the serpent in the desert, so must the Son of Man be lifted up, that all who believe may have eternal life in him. Yes, God so loved the world that he gave his only Son, that whoever believes in him may not die but may have eternal life. God did not send the Son into the world to condemn the world may be saved through him. Whoever believes in Him avoid condemnation, but whoever does not believe is already condemned for not believing in the name of God's only Son. The judgment of condemnation is this: The light came into the world, but men loved darkness rather than the light because their deeds were wicked. Everyone who practices evil hates the light; he does not come near it for fear his deeds will be exposed. But he who acts in truth comes into the light, to make clear that his deeds are done in God.

—The Gospel According to Saint John, 3:14–21

So, today, it is good and provident event to have this book written. I am glad, relieved, and altogether happy that it is finished! Getting it done is like throwing a heavy duffel bag off this sailor's shoulders where it had rested during a long trek across the uneven and bumpy countryside. Finally, I may toss it high onto the top bunk since it is past time for me to hit the hay. Now and tomorrow, I have many other things to do. I pray daily that my case will be reopened and that I will be fully exonerated. Still, in any case, as Robert Louis Stevenson says:

"When one door closes, another shall open."

Equally, in the same breath, I recall George Patton's words to his wife, Beatrice, in a letter written in October 1945 just after the end of World War II and just before his fatal car crash on December 9 near Heidelberg,

Germany, close to where he is now buried in Luxembourg: "If a man has done his best, what else is there?"

It is quite simple: In the end, the big shots tried to make me a cully, "one who is cheated or imposed upon; a dupe, a gull, or simpleton." This definition is taken from *A Sea of Words: A Lexicon and Companion for Patrick O'Brian's Seafaring Tales* written by Dean King with John B. Hattendorf and J. Worth Estes. (Second Edition, Henry Holt and Company, 1995, Page 161.) However, since I fought back valiantly against them for many years, those big shots who tried to bring me down, to compel me to capitulate to them, to make me an unwilling cully were not successful. And for that key success, I can only thank a watchful God.

HERE ENDETH THE LESSON!!

APPENDIX ONE

REAR ADMIRAL JOHN W. BITOFF, USN (RET.)
1911 Pierce Street
San Francisco, California 94115

April 30, 1999

The Honorable Norm Dicks
United States Congressman
6th District Washington
500 Pacific Avenue
Bremerton, WA 98337

RECEIVED

VIA FAX @ 10:46 a.m.
FRIDAY, 30 APRIL 1999
SENT TO ME BY 2ADM BITOFF

Dear Congressman Dicks:

This is in response to your letter of March 5, 1999 in which you asked me to provide you with a written account of my role in the case of LCDR Walter Fitzpatrick, USN (Retired). I regret the length of time it took for me to respond to your request, but the incident that eventually led to LCDR Fitzpatrick being tried by a court-martial occurred in 1988 and it required an enormous effort on my part to recall the details associated with the case. In addition, I retired from active service at the end of 1991 and I am no longer privy to the official files and other documents pertaining to this case.

As a matter of background, I was Commander Combat Logistics Group ONE (CLG-1) and Commander Naval Base San Francisco from January 1989 through October 1991. In addition, I was Commander Task Force 33, the operational commander for all logistics ships, including Military Sealift Command ships, in the U.S. THIRD Fleet. In my CLG-1 hat I had 15 major ships, including the USS MARS (AFS-1), and approximately 6000 officers and men.

I had close personal knowledge and frequent association with the 15 commanding officers in my Group. I met with them frequently and wrote their fitness reports. Conversely, I had little or no contact with the ship's executive officers and with the exception of one or two, I did not know them by name. I did not know LCDR Walter Fitzpatrick, Executive Officer, USS MARS, personally or by reputation. The USS MARS was a top performing ship with two exceptional commanding officers during my tenure. Both of these fine officers went on to command aircraft carriers and one of them became a flag officer. USS MARS was nominated by me for the coveted Battle Efficiency "E" award in both competitive cycles during my tour. She was considered to be the best AFS in the Pacific Fleet. It stands to reason that LCDR Fitzpatrick; the ship's executive officer (the number two officer in the ship's chain of command) played a significant role in USS MARS's achievement.

The incident that led eventually to LCDR Fitzpatrick's trial before a court-martial occurred in 1988, long before I assumed command of Combat Logistics Group ONE. The incident I am referring to concerned a group of USS MARS officers and enlistedmen and their spouses who represented the ship at the funeral for the brother of Captain Michael B. Nordeen, USN, the MARS commanding officer. The funeral took place at Arlington National Cemetery. Captain Nordeen's brother, also a Navy captain, was murdered by terrorists while serving in Greece. Funding to send the ship's representatives to the funeral came from the USS MARS Morale, Welfare and Recreation (MWR) fund. The decision to send a delegation from the ship apparently occurred after Captain Nordeen departed on emergency leave. Incidentally, this thoughtful gesture by MARS personnel was lauded at the highest echelons of the Navy, including the Chief of Naval Personnel

My predecessor, RADM Robert Tony, USN, did not brief me on the incident during the change of command process, and when later queried by me, indicated that he did not inform me because he believed it to be a minor matter. I first became aware of a possible problem with MARS MWR account when the ship became the subject of an MWR audit or "assist visit" by the Commander Surface Warfare Force, U.S. Pacific Fleet (COMSURFPAC) civilian Welfare and Recreation Management Specialist. Somewhere in the sequence of events, I also remember being informed of a telephone message on our Waste, Fraud and Abuse "hotline" that questioned the expenditures for the funeral trip. The distinction I am making here is that I did not ask for the audit, it was initiated by my immediate senior in the chain of command. The audit questioned the use MWR funds for sending a delegation from the MARS to the funeral. In addition, other expenditures were in question, including the purchase of a tent for official ceremonies and the purchase of several televisions and stereo sets for the ship. As a result of the audit, COMNAVSURFPAC directed me to conduct an inquiry to the allegations contained in the inspection report.

The next thing that happened in sequence was an Article 32 Investigation to determine if there was any real wrongdoing in this case. My recall is not complete as to the specific details that led up to the Article 32, but I believe my Chief of Staff came to see me in the company of LT Timothy W. Zeller, my Staff Judge Advocate, regarding the matter. LT Zeller was adamant that we conduct an Article 32 investigation, if for no other reason than to "cover our six o'clock" with higher authority. I concurred, hoping that it would the clear the air on this issue. I assumed the Article 32 investigation would follow normal practice and be conducted by a civilian special agent of what was then called the Naval Investigative Service or NIS.

I was extremely busy at this time dividing myself between my duties at my two primary commands and the increasing demands placed on me by my CTF 33 operational hat. In fact, I was deployed much of this time in Alaska and the Aleutian Islands for PACEX 89, the largest peacetime exercise in Pacific Fleet history. My CLG-1 staff remained behind in Oakland in the normal conduct of business while I was deployed aboard ship. Shortly after my return from deployment, the Loma Prieta earthquake struck the Bay Area and I found myself leading the Navy's massive rescue and recovery effort.

I clearly remember being surprised by how aggressive LT Zeller seemed to be about this case and specifically, LCDR Fitzpatrick's role in it. I liked Tim Zeller personally and i had complete faith and trust in him. However, it was obvious that LT Zeller saw most things in terms of black and white. On one occasion, during an informal conversation in my office, I told him that real life situations were often too complicated for purely black and white solutions and that sometimes the answer lies in shades of gray. He smiled, and said, "I guess it's my Marine Corps training." I mentioned this encounter to my Chief of Staff and he agreed with my assessment of LT Zeller and added that he was nevertheless, extremely persevering and serious in all endeavors.

When the Article 32 investigation was completed, I was surprised to find LT Zeller had conducted the investigation, rather than the NIS. I questioned my Chief of Staff on this point and I recall him telling me that LT Zeller had asked NIS for assistance, but they were unable to provide an agent to go to sea aboard USS MARS. I am not sure whether LT Zeller and/or my Chief of Staff briefed me when they provided me with the results of the Article 32, but I do remember not being terribly concerned with the seriousness of what I was being told. I specifically remember asking the following questions: Did anyone line their pockets with the MWR expenditures? Was there anything irregular regarding the purchase of the TVs and stereo equipment for the ship or did any of this equipment find its way to a crewmembers home or car? The answer to each of my questions was no.

LT Zeller remained hard over on the use of MWR funds for the crewmembers to attend the funeral. I did not agree that these were criminal acts, but rather "creative", albeit improper use of MWR funds and a modicum of poor judgement as well. Based on this information, I told LT Zeller that I would convene an Article 15 NJP (Admiral's Mast) in the case of LCDR Fitzpatrick. I would have taken the same action with CAPT Nordeen, the former commanding officer, but he had departed the area and I no longer had Article 15 jurisdiction over him. I did however, award CAPT Nordeen a Non-Punitive Letter of Instruction, citing the discrepancies noted in the COMSURFPAC MWR audit. I also directed that crewmembers that received funds for the trip to the funeral in Arlington National Cemetery are asked to return all, or as much as, they could afford, to the MARS MWR fund. I believe there was a reasonable attempt to do this, because I received a telephone call from one of the officers (a Navy Chaplain) who attended the funeral with his wife, telling me that he returned the funds and asked for my understanding on this matter.

I was not making light of the charges regarding the misappropriation of MWR funds. My training and upbringing in the Destroyer Force, where I spent most of my seagoing career, made me a "strict constructionist" regarding the proper administration of all funds that were entrusted in my care. However, my long experience revealed that the Naval Aviation community had a reputation for taking a different or more liberal view of MWR funds as opposed to appropriated funds. Many Naval Aviators took a more imaginative or creative approach to the administration of MWR funds. I do not mean to infer that funds were used in an illegal fashion from a criminal perspective, but rather giving short shrift to the MWR Regulations "fine print" as long as it enhanced crew morale. I have had personal experience with similar matters when I was a junior officer. Based on the aforementioned and the fact the commanding officer was a naval aviator, I concluded that this atmosphere existed on USS MARS and it should not be a surprise that the executive officer would reflect the captain's attitude. Therefore, I had no desire to single out LCDR Fitzpatrick for punishment.

I next received an office call from CAPT Kevin Anderson, USMC, certified as a Judge Advocate, who I believe was accompanied by LT Zeller, my Staff Judge Advocate. CAPT Anderson identified himself as LCDR Fitzpatrick's Defense Counsel and then informed me that LCDR Fitzpatrick would not accept Admiral's Mast / Article 15 NJP unless I guaranteed that if any punishment was awarded, it would be non-punitive (i.e. not go in his record). I was startled and incensed by this demand, particularly coming from an officer of the court. I made it clear to CAPT Anderson that he was not acting in the best interests of his client. I gave him a stern lecture and told him that I have had NJP authority, on and off, for almost 30 years, beginning as a Lieutenant commanding officer and that I never prejudged a case that came before me. On the contrary, I dismissed many cases at Captain's Mast because new information surfaced during the proceedings. As a matter of principle, I could not accede to CAPT Anderson's demands. I closed the meeting by telling CAPT Anderson, in the strongest possible terms, that he and LCDR Fitzpatrick were making a serious mistake that could have terrible consequences. I instructed him to advise LCDR Fitzpatrick that Article 15/ NJP was in his best interest. LCDR Fitzpatrick, through his Defense Counsel, chose trial by court-martial vice Article 15 /NJP. LCDR Fitzpatrick's refusal to accept Article 15/NJP left me with no legal recourse but to convene a Special Court-Martial. The court-martial convicted LCDR Fitzpatrick of violating Article 92 of the Uniform Code of Military Justice, being derelict in the performance of his duties regarding the administration and expenditure of MWR funds. Therefore, I awarded him a Letter of Reprimand.

Congressman Dicks, I tried to avoid a court-martial in this case at every turn in the road. Based on my assessment of the charges, I believed that a court-martial would be a waste of the Navy's time and money and it would unfairly single out LCDR Fitzpatrick for punishment. There was no doubt in my mind that MWR funds were used improperly and there was sufficient blame to go around. I was convinced that there was no personal gain from the misuse of these funds and in the final analysis, the ship and the Navy were the ultimate beneficiaries. However, rules and regulations are there for good reason and I, in good conscience, could not sweep the matter under the rug. While I indicated earlier that I never prejudged an Article 15 case, I necessarily went into the proceedings with a general idea or window of possible punishment if no additional information or extenuating circumstances were presented. In this case, if LCDR Fitzpatrick had accepted Article 15/NJP and nothing more untoward came out, I was prepared to award him a Non-Punitive Letter of Instruction, the same punishment that was meted out to his commanding officer. This would have allowed him to go on with his career without impediment.

I have never understood why LCDR Fitzpatrick and/or his defense counsel refused my offer of Article 15. At the time, I surmised that it was a combination of LCDR Fitzpatrick acting in a fit of peak and incompetence on the part of CAPT Anderson, his defense counsel. I distinctly remember being unimpressed with CAPT Anderson, beginning with the encounter in my office the day he refused to accept the Article 15

APPENDIX ONE

for his client. I fervently hoped that LCDR Fitzpatrick and CAPT Anderson would come to their senses before risking it all at a court-martial. Had they done so up to the very minute before the court-martial opened, I would have gladly reverted to Article 15/NJP. In addition, I was a very accessible flag officer with a career-long reputation for championing the underdog and for my friendly demeanor. Why was it that LCDR Fitzpatrick, or anyone else in his camp, did not attempt to meet with me and have reason prevail in this case?

You should be aware, that in the wake of the court-martial conviction, LCDR Fitzpatrick did exercise the right of appeal and I denied it. Frankly, by my very nature, I was inclined to grant the appeal, but after much soul searching and seeking independent opinion from other senior officers not associated with the case, I found myself with a moral dilemma. I believed the punishment awarded by the court-martial to be too harsh and that LCDR Fitzpatrick was bearing full responsibility for the events on USS MARS, but I brought the charges and I convened the court-martial in the proper conduct of my duties. How could I now throw it all to the wind just because I was not happy with the results of the proceedings that I instituted. That court was made up of a jury of his peers, who unfortunately did not see the situation in the same light, which I did. They came close however and cleared him of all charges and specifications but one. It was a tough call and I made my decision after much thought and deliberation. I have always been extremely unhappy with the outcome of this case and I wish I could have prevented the irrational behavior that brought it about.

I had to make a statement on this case in 1994 when the Chief of Naval Operations directed the Judge Advocate General of the Navy to look into the matter following a series of newspaper articles that appeared in your home state of Washington. I had hoped that upon review, the case would have been thrown out on some technicality. However, my position has not changed in the ensuing years: (1) the charges stemming from the events aboard USS MARS should not have been consummated in a court-martial; (2) LCDR Fitzpatrick was not properly served by his defense counsel; (3) LCDR Fitzpatrick, by virtue of his rank and experience, should have known that it was in his best interest to accept Article 15/NJP; and (4) LCDR Fitzpatrick should have done everything in his power to meet with me before the die was cast.

LCDR Fitzpatrick has petitioned me over the years and most recently and most ardently, during the last two months, to change all that has happened and "restore him to his rightful place on active duty." I have neither the power nor the authority to grant his wish. However, he has made a series of new and disturbing allegations, which if true, bear looking into. The most serious of these allegations from my perspective, are:

- That CAPT Anderson did not apprise him of my comments during the meeting in which he, on behalf of LCDR Fitzpatrick, refused Article 15/NJP.

• That there are serious doubts as to the validity of LCDR Fitzpatrick's signature on the Response to Letter of Reprimand dated July 17, 1990, or the so called "confession".

I am not sure what action you are inclined to take on behalf of LCDR Fitzpatrick, if any, after reading my statement. I have tried to paint a picture of the events of this case and how they unfolded to the best of my recollection. I can assure you that I carried out my important responsibilities in this case to the best of my ability. I am not a lawyer and I do not presume to have in-depth knowledge of the arcane language associated with the legal documents, procedures and other minute details associated cases of this type. I followed the advice of my Staff Judge Advocate, as well as the advice of other legal authorities throughout these proceedings. If there were procedural errors made by me, they were not intentional and the Office of the Judge Advocate General of the Navy or other competent authority has heretofore not brought them to my attention.

I believe the only option open to you to bring some humane closure to this tragedy, is to convince the Navy to review this case again in light of the troubling allegations mentioned above. Possibly, in a gesture of magnanimity, the Secretary of the Navy might grant clemency and remove the Federal conviction from his record. As an aside, but of great immediate importance, LCDR Fitzpatrick informed me that there is a move afoot to remove his security clearances as a delayed result of his long ago conviction. If this is allowed to happen it will, in all likelihood, deprive him of his ability to earn a living. Based on the circumstances of this case, removing his security clearances is not appropriate and is draconian by any civilized rule of measure.

Sincerely,

John W. Bioff
Rear Admiral, U.S. Navy (Retired)

APPENDIX TWO

**Navy Captain
William Edward Nordeen**

BY FRANK JOHNSTON—THE WASHINGTON POST

REMEMBRANCE

A funeral was held yesterday at Arlington Cemetery for Capt. William E. Nordeen, who was killed by a car bomb last week in Athens. A flag was given to his mother, while his wife and daughter watched. Story on Page A16.

APPENDIX THREE

DEPARTMENT OF THE NAVY
COMMANDER COMBAT LOGISTICS GROUP ONE
FPO SAN FRANCISCO 96601-5309

IN REPLY REFER TO:

5800
Ser N14/ 1471
2 0 OCT 1989

FOR OFFICIAL USE ONLY

From: Commander, Combat Logistics Group 1
To: Commander, Naval Surface Force, U.S. Pacific Fleet

Subj: INTERIM HOTLINE COMPLAINT REPORT, CNSP I & E 05-89

Ref: (a) COMNAVSURFPAC ltr Ser 006/9002 of 15 September 1989

Encl: (1) Hotline Interim Completion Report as of 23 October
1989

1. Enclosure (1) is provided in response to reference (a).
Enclosure (1) should be considered "raw data," reflecting
specifically the findings, opinions and recommendations of the
investigating official alone, without editing by higher
authority.

2. In view of the apparent lack of proper management associated
with the USS MARS (AFS 1) Morale Welfare Recreation (MWR) Fund
and apparent serious irregularities identified by the
investigating official, the following actions are being taken:

a. The new USS MARS (AFS 1) Commanding Officer, Captain W.
W. Pickavance, has been directed by message to secure in his
possession all MWR records for the time period of investigation
interest, and to ensure the availability of witnesses upon USS
MARS (AFS 1) return from PACEX operations.

b. Charges are being prepared in the case of Lieutenant
Commander Walter F. Fitzpartick, former USS MARS (AFS 1)
Executive Officer, and will be referred to an Article 32 hearing
appointed by this command.

c. Lieutenant Commander Walter F. Fitzpatrick is currently
in execution of PCS orders between USS MARS (AFS 1) and Naval
War College. In order to ensure proper jurisdiction is
maintained in this case, this command requested NAVMILPERSCOM
modify the PCS orders to reflect assignment of Lieutenant
Commander Walter F. Fitzpatrick to COMLOGGRU ONE on a TEMDU
FURASPERS basis, to remain in effect until resolution of this
matter. Lieutenant Commander Walter F. Fitzpatrick was verbally
notified of this action on 13 October 1989 to preclude his moving
out of his permanent residence in the San Francisco Bay area.

FOR OFFICIAL USE ONLY

ENCLOSURE (2)

281

FOR OFFICIAL USE ONLY

Subj: INTERIM HOTLINE COMPLAINT REPORT, CNSP I & E 05-89

3.   Appropriate decisions with regard to disposition of charges and conclusions of accountability will be made upon completion of the Article 32 hearing.

M. B. EDWARDS
Chief of Staff

2

HOTLINE INTERIM COMPLETION
REPORT AS OF 23 OCT 1989

1. <u>Name of Investigating Official</u>: Lieutenant Timothy W.
Zeller, JAGC, USNR.

2. <u>Billet and Address of Investigating Official</u>: Staff Judge
Advocate, Commander, Combat Logistics Group 1, NSC Oakland, Ca.
(415) 466-6125/AVN 836-6125.

3. <u>Hotline/Integrity and Efficiency Control Number</u>: CNSP 05-89.

4. <u>Allegations Investigated</u>: Abuse of monies from the Morale,
Recreation and Welfare Fund, particularly the expenditure of
funds to send certain members of the USS MARS (AFS 1) and spouses
to a funeral and the expenditure of funds by sending two members
to Hawaii for an alleged MWR brief. The investigation was
broadened in accordance with regulations to include all other
wrongdoing(s) discovered in the expenditure of MWR funds.

5. <u>Evidence Examined</u>:

    a. CNSP Audit report of 1 Sep 1989
    b. NAVMILPERSCOMINST 1710.3A
    c. BUPERSINST 1710.11A
    d. Interview of CAPT Michael B. Nordeen, previous
       Commanding Officer, USS MARS (AFS 1)
    e. Interview of CDR T. A. Rorex, Senior Supply Officer USS
       MARS (AFS 1)
    f. Interview of LCDR W. F. Fitzpatrick, Executive Officer
    g. Interview of LT B. Ableson, CHC
    h. Interview of LT J. Samples, current MWR Fund Custodian
    i. Interview of LTJG L. D. Vaughn, with receipts for trip to
       funeral
    j. Interview of HMC M. W. Collins, Rec Committee Member
    k. Interview of SKC G. F. Esposto
    l. Fiscal year 1988 MWR Report
    m. Custody Cards for Electronic equipment purchased by MWR
    n. USS MARS Instruction 1710 dated 1985
    o. USS MARS Instruction 1710 (Proposed)
    p. Copies of all available checks and bank statements
    q. Proposed Fiscal year 1989 MWR Report
    r. Fiscal Year 1989 Recreation Committee minutes
    s. Interview of SKC L. N. Strong, Current MWR Director
    t. Interview of SK3 E. D. Brown, Rec Committee Member

6. <u>Circumstances and Facts</u>:

Out of the $100,000.00 expended from the MWR Fund during
Fiscal Year 88, it is apparent that approximately twenty percent

ENCLOSURE (9) ~~ENCLOSURE (1)~~

was misspent. This figure does not include the cost of the hail and farewell, since this expenditure actually was paid for in FY 89.

SPECIFIC INSTANCES OF IMPROPER EXPENDITURES

1. Funeral Party.

a. On or about 1 July 1988, Commanding Officer, USS MARS (AFS-1) received a telegram stating that his brother had been murdered by terrorists. CAPT M. B. Edwards, Assistant Chief of Staff at CLG-1, was immediately dispatched to SOCAL. At the time of CAPT Edwards' arrival to temporarily relieve CAPT Nordeen, USS MARS (AFS-1) was engaged in REFTRA in the SOCAL OP area. Turnover lasted approximately one hour, after which CAPT Nordeen departed the area by helo.

b. That day, the Chaplain, LT Ableson, was put ashore to observe CACO assistance for CAPT Nordeen's sister-in-law. Upon returning, he was told by the Executive Officer that some of ships' personnel would be attending the funeral. The Chaplain indicated that the appropriate leader would be line officer. The Executive Officer subsequently sent the Chaplain as the senior member.

c. Prior to departure of the team (which consisted of two Officers, the Command Master Chief, five Enlisted Personnel and the spouses of the Executive Officer, Chaplain, Doctor and a Master Chief), the Master Chief called a meeting of the Recreation Committee, whose actions are advisory in nature, voted affirmatively for sending military personnel and flowers, but voted unanimously against paying for spouses. According to one witness, the implication from the Master Chief was that the committee would either go along or would be on the "shit list". The personnel in the funeral party were unaware of the vote not to send spouses.

d. The decision to send the party, including the spouses, lies with the Executive Officer. The Executive Officer stated that after the MWR meeting, he held a meeting on the fantail of all crewmembers. The content of the talk given by the Executive Officer differs between the story of the Executive Officer and the other members involved. The Executive Officer gives the impression that he stated that sending the military members and the spouses had been approved by MWR, but that he wanted anyone that had an objection to the expenses being paid by MWR to get word to him. The other version of the story relates that there was no mention of the spouses at all, and that the implication was that objections would have to be voiced at that moment on the fantail. One of the crewmembers relates that it was even

2

presented that the Executive Officer would pay for the trip himself if the crew did not approve, but that either way the crewmembers were going.

    f. At the time of the decision, the Executive Officer was not the acting Commanding Officer. Evidence indicates that the temporary Commanding Officer, CAPT Edwards, was only aware that a party of crewmembers were attending the funeral, without being advised how it was being paid for or that spouses were included.

    g. CAPT Nordeen was unaware the MWR funds had been used to pay any expenses of the trip until 2 or 3 months later. Even then he was not aware that the spouses' tickets had been paid for with MWR Funds.

2. Hawaii Trip.

    a. OSC Wagoner received a check for $1400.00 to fund a trip for himself and LT Dorris to Hawaii for an MWR/Operations brief. It is interesting to note that LT Dorris, the Operations Officer, had no connection with MWR other than Athletic Director.

    b. The Executive Officer disclaims any knowledge of the fact that an OPS Brief was taking place at the same time as the trip. The check in this case was signed personally by the Executive Officer. The Commanding Officer, CAPT Nordeen, stated that even though he knew the trip was to be dual purpose, MWR and OPS Briefs, he did not know until later that MWR funds had been used to pay for the trip.

    c. There is no evidence at the present time that any MWR brief was ever scheduled or took place in Hawaii.

3. Electronic Equipment Expenditures

    a. This abuse of funds by the Executive Officer relates to purchases of equipment (stereo's, televisions and video recorders) for exclusive use by the Commanding Officer, Executive Officer and the Command Master Chief.

    b. Prohibitions against MWR funds being used for such purposes are contained in NAVMILPERSCOMINST 1710.3A and BUPERSINST 1710.11A, as well as in the USS MARS Instruction governing such funds. The impropriety of the acquisition was pointed out to the Executive Officer at the time of the purchase and afterward by LCDR Dolan, the Assistant Supply Officer.

    c. All purchases were authorized the Commanding Officer by a general statement that he wanted to upgrade the gear onboard. The equipment was picked out and purchased by the Executive

3

Officer with MWR Funds. It is the contention of the Executive
Officer that these funds were properly spent due to his belief
that they were part of the crew also.

d. The instructions clearly prohibit the expenditures of
funds if the benefit will only be for a few, in this case, only
one.

e. The electronics' bill from this mass purchase amounted to
approximately $6500.00.

4. Hail and Farewell

a. Although the majority of the problems addressed occurred
in 1988, the problem continues. A recent Hail and Farewell for
the departing and oncoming Commanding Officers was paid for to a
large extent out of MWR funds.

b. The matter was brought up before the MWR committee, which
agreed to fund the event up to $2,000.00, provided the entire
crew was invited. The fact of the situation was simply that the
additional cost of the outing ($60.00 per person) was such that
few enlisted personnel could have afforded it. It was also
apparent that even though the sign up list was readily available
to the officers and chiefs, the same was not true for the
enlisted personnel in paygrades E-6 and below.

5. Promotional Items.

a. The MWR-funds are spent to fund minor items of promotion
for the ship as well. USS MARS (AFS 1) in the practice of
distributing Mars candy bars to visiting VIPs, visiting CO's and
others, nicely packaged on a miniature pallet. These items are
paid for out of MWR.

b. There is some indication that this cost is being
reimbursed.

OVERSIGHT PROBLEMS

1. No direct access of the Fund Custodian to the Commanding
Officer.

a. LT Samples has been required to go through the Supply
Officer and the Executive Officer to obtain direction. No
personal access was provided to the Commanding Officer.

2. Failure to control preprinted MWR Checks.

a. Current regulations dictate that a tight control be kept
on all preprinted checks and a strict accounting be maintained.

4

b.   The Executive Officer was in the habit of taking several checks at a time to use for various items without explanation or receipts.

c.   Several checks which were taken have shown up on the bank statements but were never actually returned to the custodian.

d.   All bank statements go through the ship's office prior to being sent to the Supply Department for the fund custodian.   It has been known that sometimes the statements have been open prior to being received by the fund custodian.

e.   The Executive Officer states that he never saw the Bank Statements.

3.   Failure to maintain Records and submit reports..

a.   All records and receipts for FY 88 and prior are missing with the exception of some cancelled checks and a rough copy of the FY 88 report.   The later was recovered from the Executive Officer during the time I was on board.   It had never been forwarded to CNSP or NMPC-65.

b.   The missing records were discovered upon the departure of OSC Wagoner.

c.   During the interview with the Executive Officer, LCDR Fitzpatrick claimed that he was unaware of the requirement to send the reports to NMPC and CNSP. However, when the investigating officer obtained a copy of the FY 88 report from the Executive Officer, attached to the report were two messages from CNSP, both of which outlined the proper procedures and addressees for the report.   The messages had a date time group of 16 and 20 September 1988, respectively.

7. CONCLUSIONS OF THE INVESTIGATING OFFICER

CAPT Nordeen is guilty of dereliction of duty by failing to account for the proper expenditure of MWR Funds.

LCDR Fitzpatrick is guilty of dereliction of duty by failing to adhere to proper procedures for the expenditure of MWR funds, violation of the Standards of Conduct by using his authority with MWR funds for his own aggrandisement and several counts of larceny due to the diversion of monies to personnel not attached to the crew, including his spouse.

5

APPENDIX FOUR

PERSONAL FOR RADM BITOFF

Dear Admiral,

Prior to your final consideration of the disposition of the case against _____ , I feel obligated to communicate some reservations in regard to the absence of disciplinary measures. The one question that I received from almost all concerned, including the members, was why wasn't it _____ on trial instead of LCDR Fitzpatrick. After the trial, the senior member of the board was left with the distinct impression that _____ was looking down on the entire court-martial knowing that they could not touch him. I recognize the political ramifications of pressing the case against _____ , but I also recognize that he could have prevented the entire problem if he had performed his job as he should, especially when he found out about the Hawaii Trip a few days after it happened.

Since I don't believe in keeping a file to cover this office when decisions are later questioned, there is no copy of this letter in my files or on my computer.

Very Respectfully,

T. W. Zeller

RECEIVED BY MAIL

Saturday, 18 July 2001

Walter Francis Fitzpatrick, TD.

288

APPENDIX FIVE

NAVY HOTLINE PROGRESS REPORT
AS OF 4 OCTOBER 1989

1. Applicable DON Organization: Commander, Combat Logistics Group 1, NSC Oakland, Ca.

2. Hotline Control No.: CNSP I&E 05-89

3. Date Referral Initially Received: 15 September 1989

4. Status: The investigation has revealed that approximately twenty percent of the 100,000 dollars expended from the MWR fund in Fiscal Year 1988 was misspent. Due to the wide dispesion of the personnel involved, the accountability issues are still being addressed. Although the current regulations require the inclusion of the Naval Investigative Service in an investigation when possible wrongdoing has been discovered, the deployment of the USS MARS has made such action impractical. The investigating officer embarked on the ship to conduct the investigation.

5. Date of Expected Completion: 15 October 1989.

6. Action Agency Point of Contact:

Enclosure (3)

APPENDIX SIX

RECEIVED *BY MAIL*

Saturday 28 July 2001

2 NOV 89

From: N14
To: 02

Subj: Chronology of events in regard to LCDR Fitzpatrick

Sir, the following are submitted for your consideration.

1. Prior to 31 August 1989, CNSP had tried to schedule the audit but was stalled by LCDR Fitzpatrick. The exact date of the original audit date cannot be obtained until Ms. Christopherson returns to work on Monday.

2. 6 October 1989. LCDR Fitzpatrick visited Lt Zeller's Office on his own volition, bringing with him several personal records and logbooks, some of which dealt with MWR activities. LCDR Fitzpatrick was advised at that time that he was still under no obligation to talk and that his rights under Article 31 could still be invoked.

3. The letter forwarding my report to CNSP pointed out clearly that the report was unedited and the conclusions reached were mine alone. Your entry for 5 Oct 89 may call that statement into question, should this document ever be needed. This issue is important to keep our roles separate in this case, namely I do not want you disqualified from action later in the case for being an accuser. Recommend you drop the second sentence.

Very Respectfully,

T. W. Zeller

RECEIVED *By mail*

Saturday 28 July 01

```
5800
N-02
03 Nov 89
```

MEMORANDUM FOR THE RECORD

Subj: LCDR W. F. FITZPATRICK, USN, 551-90-4692/1110

Ref: (a) COMNAVSURFPAC ltr 5041 Ser 006/9002 of 15 Sep 89
     (b) LCDR Fitzpatrick ltr of 19 Oct 89 to SECNAV
     (c) CHNAVPERS Washington DC 240539Z OCT 89 (Bupers ORD 1399(03))

1. The following chronology pertains to the decision to initiate an Article 32 investigation and hearing ICO SNO.

2. Chronology of Events

31 AUG 89   CAPT W.W. Pickavance relieves CAPT M.B. Nordeen as C.O. USS MARS (AFS 1).

01 SEP 9   COMNAVSURFPAC Welfare and Recreation Management Specialist R. A. Christopherson conducts Welfare and Recreation inspection of MARS. Fund adjudged to be unsatisfactory. Multiple discrepancies discovered.

15 SEP 89   COMNAVSURFPAC directs COMLOGGRU ONE (CLG-1) to conduct Integrity and Efficiency investigation into alleged misuse of Welfare and Recreation funds aboard MARS as discovered during 01 Sep 89 inspection. Reference (a) pertains.

15 SEP 89   C.O. MARS advises COS, CLG-1 (CAPT Edwards) of intent to have newly arrived PXO MARS relieve LCDR Fitzpatrick during first phase of PACEX 89 and detach LCDR Fitzpatrick at first Aleutians portcall. Asks if LCDR Fitzpatrick can come TAD to CLG-1 to study for/take command qual exam, then go on leave pursuant to executing PCS orders to NAVWARCOL Newport. COS CLG-1 agrees.

18 SEP 89   LT Zeller CLG-1 SJA, embarks MARS to conduct I&E investigation. MARS departs Oakland, CA for PACEX 89.

29 SEP 89   LCDR Fitzpatrick reports to CLG-1 after detaching MARS in Alaska. Advises acting COS (CAPT Romanski) that he does not desire to take command qual exam (states "C.O. MARS must have misunderstood."), desires to go on leave, will take exam some other time while in Newport. Further advises his personal plans are to execute "Do It Yourself Move" and that he/family will depart local area end-October/early November to drive to Newport. Indicates he is having trouble getting Navy housing commitment from NAVWARCOL. Told by CAPT Romanski that NWC treats everyone that way and that he should press for a commitment.

Subj: LCDR W. F. FITZPATRICK, USN, 551-90-4692/1110

Note: For clarity, CLG-1 (RADM Bitoff) departed San Francisco area on 17 Sep 89 for PACEX and returned 4 Oct 89. COS CLG-1 (CAPT Edwards) departed 17 Sep 89 and returned CLG-1 on 10 Oct 89.

| 02 OCT 89 | LT Zeller returned from MARS embark, verbally debriefed CAPT Romanski that he felt case against LCDR Fitzpatrick was serious. Told to complete (b)(5) initial written report (extensive) for review by CAPT Edwards/Romanski, then presentation to RADM Bitoff. |

05 OCT 89   LT Zeller presents initial report draft to CAPT Romanski.

06 OCT 89   LCDR Fitzpatrick visited LT Zeller's Office on his own volition, bringing with him several personal records and logbooks, some of which dealt with MWR activities. LCDR Fitzpatrick was advised at that time that he was still under no obligation to talk and that his rights under Article 31 could still be invoked.

10 OCT 89   CAPT Edwards returns to CLG-1. He and CAPT Romanski review LT Zeller's investigation draft.

11 OCT 89   RADM Bitoff completes Fleet Week activities which commenced 05 Oct.

12 OCT 89   RADM Bitoff meets with CAPT Edwards/Romanski and LT Zeller to discuss MARS Wel/Rec investigation and LCDR Fitzpatrick. RADM Bitoff approves recommendation to refer matter to Article 32 to ensure complete impartiality, and to have LCDR Fitzpatrick's PCS orders modified to reflect TEMDU FURASPERS at CLG-1 (vice transfer to NAVWARCOL) to retain CLG-1 jurisdiction.

12 OCT 89   CAPT Romanski calls NMPC-411 and requests ORDMOD for LCDR Fitzpatrick, explains reason.

13 OCT 89   CAPT Romanski calls NMPC-411 again and confirms that ORDMOD will be issued.

13 OCT 89   CAPT Romanski calls LCDR Fitzpatrick (at home in Clayton, CA); advises that Welfare and Rec investigation is being referred to Article 32 and that PCS orders will be modified. LCDR Fitzpatrick responds:

              - Entire matter is a "kangaroo court."

2

Subj: LCDR W. F. FITZPATRICK, USN, 551-90-4692/1110

> - Entire matter is unfair, immoral and poor leadership because he, LCDR Fitzpatrick, should have been told well before 13 Oct of the possibility of his orders being changed. Advised he had already started getting ready to move out of Clayton residence and had mailed deposit for a Newport rental.
>
> - Requested in strongest possible terms that orders to Newport be reinstated. Told that would not be done by CLG-1.

16 OCT 89   Defense counsel from NLSO Treasure Island assigned LCDR Fitzpatrick.

19 OCT 89   LCDR Fitzpatrick writes letter to SECNAV requesting reinstatement of orders to NAVWARCOL. Reference (b) pertains.

26 OCT 89   CLG-1 receives requested ORDMOD from NMPC. Reference (c) pertains.

27 OCT 89   COS CLG-1 receives copy of reference (b) by certified mail.

P. A. ROMANSKI
CAPT   USN

3

APPENDIX SEVEN

OPNAV 5216/144C (8-81)
S/N 0107-LF-062-2324

DEPARTMENT OF THE NAVY

Memorandum

DATE: 9 NOV 89

FROM: N14

TO: OO
VIA: 01 [signature] Concur   02 [signature] Concur

SUBJ: Grant of Immunity.

Sir, Enclosed is a grant of immunity
for the former fund custodian
of Mess NWR. As a practical
matter, the navy lost jurisdiction
over former LT feeley when
he was released from active
duty. Nonetheless, he has requested
immunity prior to answering
questions from the government
or defense. To ensure fairness
to the accused and the government, I recommend the grant
be given.

[signature]   (b)(5)

RECEIVED *Via Mail*
Saturday 28 July 2001

294

Date: 9 November 1989*

From: Lieutenant Timothy W. Zeller (CLG** – 1 Staff JAG Code N14)

To: Rear Admiral John W. Bitoff (Commander CLG – 1 Code 00)

Via: Captain A.E. Millis (Bitoff's chief of staff Code 01) (Concur)
     Captain Paul Romanski (assistant chief of staff Code 02) (Concur)

Subject: GRANT OF IMMUNITY

Sir, Enclosed is a grant of immunity for the former fund custodian of MARS [Morale, Welfare and Recreation] MWR.

As a practical matter, the Navy lost jurisdiction over former Lieutenant Feeley when he was released from active duty.

Nonetheless, he has requested immunity prior to answering questions from the government or defense.

To ensure fairness to [Lieutenant Commander Fitzpatrick] the accused and the government, I recommend the grant be given.

                        Very respectfully,
                        /s/
                        Timothy W. Zeller

Notes:
- *This CLG – 1 internal memo was secreted from 9 November 1989 until Saturday, 28 July 2001 – nearly 12-years)

- **CLG – 1 : Combat Logistics Group ONE
    - Admiral John Bitoff – Commander (Code 00) (Bitoff's signature appears on the immunity grant)
    - Captain Millis – Chief of Staff
    - Captain Romanski – Assistant Chief of Staff (Code 02)
    - Lieutenant Timothy W. Zeller – Staff Judge Advocate General (Bitoff's staff JAG – Bitoff's military attorney staff assistant)

IN THE MATTER OF   )
)
U. S. v FITZPATRICK   )
)
)

14 NOV 1989

GRANT OF IMMUNITY

To: Mr. Brian J. Feeley,   ., Thompkins County, (b)(6) &
New York

1. It appears that you are a material witness for the Government in the matter of LCDR Walter Francis Fitzpatrick, U. S. Navy, 551-90-4692/1110. Charges have been preferred to an Article 32 Investigation for the following alleged offenses.

    (a) Violation of the UCMJ, Articles 92, 108, and 121.

        (1) Article 92: Standards of Conduct and Dereliction of Duty.

        (2) Article 108: Suffering waste of U. S. Government property.

        (3) Article 121: Wrongful Appropriation of Morale, Welfare, and Recreation property and Misappropriation of Morale, Welfare, and Recreation Funds.

2. In consideration of your testimony as a witness for the Government in the foregoing matter, you are hereby granted immunity from prosecution for any offense or offenses arising out of the matters therein involved concerning which you may be required to testify under oath.

3. It is understood that this grant of immunity from prosecution is effective only upon the condition that you actually testify as a witness for the Government. It is further understood that this grant of immunity from prosecution extends only to the offense or offenses in which you were implicated in the matter herein set forth and concerning which you testify under oath.

J. W. BITOFF
Rear Admiral, U. S. Navy
Commander
Combat Logistics Group 1

APPENDIX EIGHT

Memorandum
11 January 90

From: N14
To: 00
Via: 01

Subj: Article 32 ICO LCDR Fitzpatrick

Sir, the Investigating Officer has recommended that disciplinary action be taken in the case of LCDR Fitzpatrick for violation of Articles 92, dereliction of duty, Article 108, Willfull suffering of loss of Government Property and Article 92, Disobeying a general order, i.e. Standards of Conduct in the personal use of vehicle.

(b)(5)

The recommendation by the Investigating Officer is qualified. On the one hand he states that the offenses are too serious to ignore, yet on the other hand he recommends Admirals Mast due to the fact that he believes there was no self-enrichment by LCDR Fitzpatrick and the fact that the IO considers the offenses to be of a pure military nature. The investigating officer's recommendation is just that, you may choose to accept it or reject it as you please.

Options: At the present time the Article 34 advice is being prepared in the event you desire to convene a general court martial. In the event you desire to do otherwise, the following are less severe options:

b)(5)

NJP, punishment imposable is restriction, Arrest in Quarters, forfieture of one-half pay times two months, and a Letter of reprimand. If you elect to impose NJP, I recommend a letter of reprimand and forfieture of pay, the latter since the offense is essentially fiscal irresponsibility. Board of Inquiry possible.

Special Court Martial, the penalties are basically the same as for NJP, the main difference is that the SPCM conviction would undoubtedly lead to a Board of Inquiry for misconduct.

(b)(5)

Administrative matters: Due to the fact that the LCDR is no longer attached to the ship, a DFC is not feasible.

Very Respectfully,

T. W. Zeller

RECEIVED
Saturday, 18 July 2001
VIA REGULAR MAIL

297

APPENDIX NINE

MEMORANDUM   31 MAY 1990

I concur 6/1

From: 006
To: 00
Via: 01 ___ 02 _____

Subj: Convening Authority Action ICO LCDR Fitzpatrick

Sir, enclosed are the action and the Letter of Reprimand ordered
awarded by the court members in the subject case. Also enclosed
is a clemency request from the defense counsel in which he
recommends that you disapprove the findings of the court,
essentially overturning the court-martial, based on his opinion
that a court was not the proper forum. This contention is
somewhat ironic in view of the fact that the accused was offered
a fair hearing at mast and refused that opportunity. I strongly
recommend that clemency not be granted, and that the sentence of
the court-martial be carried out as adjudged. Your execution of
the action and the letter will execute the sentence.

                                   Very Respectfully,

                                   T. W. Keller

RECEIVED

0936 WEDNESDAY, 11 OCT 2000

298

APPENDIX TEN

ATTORNEY WORK PRODUCT

23 Nov 89

MEMORANDUM

From: N14
To: 00
Via: 01_____02

Subj: Government Counsel Performance ICO LCDR Fitzpatrick

Sir, in the past week it has become apparent that we are not receiving the appropriate service from the Government Counsel in the Fitzpatrick case. LT Becoehian has repeatedly refused to repeat our position on witnesses, and seems to be willing to give the defense counsel anything and everything that he desires.

I instructed the GC that we would be willing to produce three witnesses that are necessarily involved to the extent that a phone call would never suffice. He was further instructed that any additional witnesses would have to be ordered by the investigating officer with the appropriate consideration given to the time required to present such witnesses. It was understood that he would do everything possible to provide the substitutes for live testimony that are allowed for in RCM 405. None of these efforts have been made by the government, with the case of Capt Edwards being a perfect example. Capt Edwards will undoubtedly be required to attend a hearing on this Saturday while he is in town. He has never been contacted by either counsel, and the testimony he will be asked for is of such minor duration that he will spend more time driving than on the witness stand.

A second example is the appearance of Ms. Ruth Christopherson, the MWR coordinator that conducted the audit. The GC assumed that she would be able to come down without any effort to contact her or this office in regard to a formal witness request. The niavete of the GC is apparent when he states that the witness can be compelled to come even though the GC and the DC have not done anything to discover what she will testify to. The GC cannot get it through his head why my one star should not go to a three star and tell his to produce a witness because the time of the lawyers is more valueble than that of the witnesses.

Notwithstanding the obvious inexperience of the Government counsel, I am sincerely convinced that the GC does not have the desire to put the effort into this case which will be required. An example of this is the fact that he will not be present when Capt Edwards is called and does not see any reason for there to be a substitute GC. Evidently the GC has alternate plans for this weekend and assumes the Investigating officer will do his job for him. Although there is no requirement in the MCM that a GC be appointed at all, the complexity of this case requires the dedication of someone who desires to win. We asked for an above average counsel for the Defense in order to ensure that the trial be fair, and for a military Judge to ensure that the complexity of the case will be appreciated. Due to the command influence factor, we specifically did not ask for a certain GC.

Unfortunately it seems as though the one we were assigned lacks not
only experience, but also desire. One can be overcome by the other,
but the absence of both leads to an untenable position.

    With regret, it is recommended that corrective action be taken
immediately to assign a special prosecutor to this case that will give
it the attention it merits. This action is required if we are to use
this hearing to find out all the facts of the improprieties alleged,
with the alternative being that the case may well be seriously
jeopardized for lack competent representation.

                    Very Respectfully,

                    T. W. Zeller

APPENDIX ELEVEN

My name is Matthew K. Bogoshian. I am currently a civilian attorney practicing law in Anaheim, California. I desire to make the following statement concerning my assignment as prosecuting attorney in United States v. LCDR Walter F. Fitzpatrick, III, USN, 551-90-4692. At that time I was a U.S. Navy Judge Advocate Corps Lieutenant, on active duty, and assigned to the Naval Legal Service Office in Treasure Island, California.

In addition to being the prosecutor at LCDR Fitzpatrick's Special Court-Martial, I was also the attorney representing the government at the Article 32 Investigation that preceded the trial.

As the prosecuting attorney in the case mentioned above and the attorney for the government at the Article 32 Investigation, I had an opportunity to work with and observe LT Timothy W. Zeller. LT Zeller was the Staff Judge Advocate (SJA) to RADM John W. Bitoff, Commander, Combat Logistics Group ONE (COMLOGGRU-1). COMLOGGRU-1 was the Convening Authority for LCDR Fitzpatrick's case.

LT Zeller was difficult to work with on this case and I did not enjoy the experience. LT Zeller seemed obsessed with the prosecution of LCDR Fitzpatrick. It was my impression that LT Zeller had a gut feeling, correct or not, that LCDR Fitzpatrick was a bad egg, and LT Zeller was intent on doing everything he could to show that. LT Zeller was a real pit bull on LCDR Fitzpatrick's case.

I remember that LT Zeller was LCDR Fitzpatrick's accuser and that a tremendous number of charges were preferred to the Article 32 Investigation. From my own research as prosecuting attorney, as borne out by the Article 32 Investigation and Special Court-Martial, the majority of charges LT Zeller brought against LCDR Fitzpatrick seemed to have little or no basis in reality, i.e., there was an absence of much if any evidence to support them. After completing my research I remember thinking that LT Zeller was quite unusual for bringing all the charges he did against LCDR Fitzpatrick.

LT Zeller ensured that I got all the witnesses I needed on LCDR Fitzpatrick's case. It is my impression that a great deal of money was spent for LCDR Fitzpatrick's prosecution.

I remember being told that there had been a Integrity and Efficiency Investigation regarding LCDR Fitzpatrick and the USS MARS (AFS-1), but I cannot remember when I learned of it.

I do remember being told that someone had contacted LCDR Fitzpatrick's promotion board (for Commander) to notify that board of the results of LCDR Fitzpatrick's trial.

The way in which LT Zeller handled this case was unusual. Also, the significant attention he gave the case was noteworthy and struck me as exceptional.

I have read and fully understand this statement, and I swear/affirm that it is true and correct.

Matthew K. Bogoshian

STATE OF CALIFORNIA)
) ss.
COUNTY OF Orange. -)

On this 8th day of April, A.D. 1992, before me, the undersigned a Notary Public in and for the State of California duly commissioned and sworn, personally appeared Matthew K. Bogoshian, to me known to be the individual described in and who executed the foregoing instrument, and acknowledged to me that he signed and sealed the said instrument as his free and voluntary act and deed for the uses and purposes therein mentioned.

WITNESS my hand and offical seal hereto affixes the day and year in this certificate above wiitten.

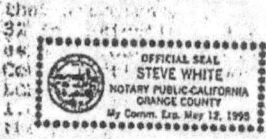

OFFICIAL SEAL
STEVE WHITE
NOTARY PUBLIC-CALIFORNIA
ORANGE COUNTY
My Comm. Exp. May 12, 1995

Notary Public in and for the
State of California
Residing at Anahem, CA

2

APPENDIX TWELVE

ZELLER'S MESSAGE TO GLENN N. GONZALEZ

FROM: Lieutenant Commander Timothy W. Zeller, Judge Advocate assigned in *USS INDEPENDENCE* (CV-62).

TO: Captain (select) Glenn N. Gonzalez, Deputy Force Judge Advocate, Commander Naval Surface Force, Pacific Fleet under Captain Richard Stewart (CNSP Force JAG, under three-star Vice Admiral David M. Bennett headquartered in San Diego, CA).

Message classification (at the time of transmission: CONFIDENTIAL! SPECIAL CATEGORY! [SPECAT]).

Time of transmission: 03:31 hours GMT, 16 April 1992

Special handling instructions and markings applied by radiomen in San Diego upon message receipt: MANUAL DISTRIBUTION REQUIRED! CNSP #2036 SPECIAL CONTROL NUMBER (SPCN) #1264. Copy 2 of only 3 copies made. Page count applied.

SUBJECT: HOTLINE COMPLAINT SUBMITTED BY LCDR FITZPATRICK, USN AGAINST [REAR ADMIRAL JOHN W. BITOFF (Commander Combat Logistics Group ONE [COMLOGGRU ONE] AND MEMBERS OF BITOFF'S STAFF TO INCLUDE BITOFF'S STAFF JAG, TIMOTHY W. ZELLER).

1.1 REMARKS (UNCLASSIFIED): THE FOLLOWING IS QUOTED FOR USE BY [CAPTAIN GLENN N. GONZALEZ] DEPUTY FORCE JUDGE ADVOCATE.

• 2. (CLASSIFIED CONFIDENTIAL AT THE TIME OF TRANSMISSION AND RECEIPT) QUOTE:

In response to [Captain Gonzalez's telephone request of 15 April 1992 for Lieutenant Commander Zeller's statement in regard to subject complaint submitted by Lieutenant Commander Fitzpatrick, USN-Ref A], the following information is provided.

No records are available on board [USS INDEPENDENCE (CV-62) homeported in Yokosuka, Japan on 16 April 1992] to refresh LCDR Zeller's memory of the incident [that occurred in 1989–1990].

In the fall of 1989, [approximately] 10 days prior to the commencement of [PACIFIC EXERCISE 1989], [Combat Logistics Group ONE] was tasked by [Commander Naval Surface Force, Pacific Fleet] to conduct an [Integrity and Efficiency] investigation into irregularities of the Morale Welfare and Recreation account of USS MARS (AFS-1).

It was known at that time that the executive officer [LCDR Fitzpatrick] was intimately involved with the administration of the [MWR] account.

[Commander, Combat Logistics Group ONE] at that time was [Rear Admiral one-star John. W] Bitoff whose Chief of Staff was Captain [Michael B.] Edwards.

The alleged irregularities involved an expenditure of approximately [ten thousand dollars] $10,000.00 to send certain members of the [USS MARS] crew to the funeral of the USS MARS CO's brother [Navy Captain William Edward Nordeen], a Naval Attaché who had been murdered [by the terrorist group known as 17 November].

[Captain Edwards, Chief of Staff to Commander, Combat Logistics Group ONE] had been detailed to replace the present CO, [Captain Michael B.] Nordeen while [Captain] Nordeen attended his brother's funeral.

At the time of the incident, USS MARS was undergoing [refresher training in San Diego, CA].

The detailing of the [Integrity and Efficiency] investigation officer was decided by [Rear Admiral] Bitoff and [Bitoff's Chief of Staff, Captain] Edwards.

Originally the plan was to send the [Combat Logistics Group ONE staff officer assigned to Bitoff as] the assistant training officer [Lieutenant Commander Steve] Letchworth, and LCDR Zeller, then a lieutenant.

Just prior to the detailing, it was decided [by Bitoff and Edwards that lieutenant] Zeller would conduct the [Integrity and Efficiency] investigation alone.

As memory serves, this decision was made for 2 reasons.

First, [Lieutenant] Zeller was one of the few officers on the [Combat Logistics Group ONE] staff who had not had a confrontation with LCDR Fitzpatrick, the executive officer.

Second, with [Combat Logistics Group ONE] immersed in [PACIFIC EXERCISE 1989], other options were limited.

These factors were considered to override normal protocol that a junior not investigate a senior.

[Lieutenant] Zeller departed onboard *USS MARS* to conduct the [Integrity and Efficiency] investigation while the ship was en route to Dutch Harbor [Alaska].

During the course of the investigation, LCDR Fitzpatrick asserted that Captain Edwards was cognizant of the plan to send [*USS MARS*] crew members to the funeral. There is no recollection that LCDR Fitzpatrick asserted Captain Edwards approved of the action.

Upon completion of the investigation [Lieutenant] Zeller departed the ship in Dutch Harbor and flew to the [Naval Shipyard] Puget Sound where the advance base of [Combat Logistics Group ONE] was located for [PACIFIC EXERCISE 1989].

An initial debriefing was delivered to RADM Bitoff, and if recollection is correct, Captain Edwards was present.

The assertion by LCDR Fitzpatrick concerning Captain Edwards's alleged acquiescence/approval of the plan for the funeral party was to the best of [Zeller's] recollection not discussed.

If memory serves, this initial debrief was cursory for two reasons.

First, though the material and interviews had been completed, they had not yet been transformed into a written report with conclusions and recommendations.

Second, the more immediate [Combat Logistics Group ONE] command concern was an unrelated officer misconduct case which had been handled by the investigating officer for another [Combat Logistics Group ONE] ship during the transit to Dutch Harbor.

Upon return of [Lieutenant] Zeller to the [Combat Logistics Group ONE] staff [headquarters in Oakland, CA], and the subsequent return of Captain Edwards, the subject of Captain Edwards's knowledge of the funeral party funding was raised.

Captain Edwards related that he was unaware of the involvement of [Morale Welfare and Recreation] funds for the funeral party until well after the incident occurred.

The assertion by LCDR Fitzpatrick that Captain Edwards was present at a ship's formation when it [MWR funding for the funeral party] was discussed was found to be incredible.

Captain Edwards was escorted to the formation to address the crew after the [Executive Officer LCDR Fitzpatrick] had completed his presentation [to crew assembled on the flight deck of the USS MARS].

Captain Edwards's presentation was in regard to the [refresher training] evolution.

As the investigating officer, [Lieutenant] Zeller determined the assertions [by LCDR Fitzpatrick] of Captain Edwards's involvement to be unbelievable.

Notwithstanding this fact, if memory serves, the subsequent decisions in the case were made by RADM Bitoff with recommendations from the investigating officer [Lieutenant Zeller] and Captain Paul Romanski, Assistant Chief of Staff for Logistics.

The issues discussed involved the disposition of charges based not only on the funeral party incident, but also numerous other irregularities in the [*USS MARS*] MWR account which were discovered during the investigation.

Subsequent to the [preferring] of charges against LCDR Fitzpatrick and prior to the final disposition of same, LCDR Fitzpatrick was offered the option of non-judicial punishment.

Upon [LCDR Fitzpatrick's] refusal, RADM Bitoff was advised of his options [by Lieutenant Zeller] of his [RADM Bitoff's] options for referral to a Special or General Court-martial.

As an Article 32 investigation had already been conducted, RADM Bitoff referred the case to a [Special Court-martial], deciding against a [General Court-martial] as [RADM Bitoff] did not desire to penalize [LCDR Fitzpatrick] for exercising his rights.

• 3. (UNCLASSIFIED) The above statement was drafted by [Lieutenant Commander] T. W. Zeller, JAGC.

Declassification instructions: Originating Agency Determination Required (OADR)

APPENDIX THIRTEEN

DEPARTMENT OF THE NAVY
PUGET SOUND NAVAL SHIPYARD
BREMERTON, WASHINGTON 98314

IN REPLY REFER TO:
5800
Ser 811/2447
02 Dec 93

FOR JUDGE ADVOCATE GENERAL'S EYES ONLY

From: Commanding Officer, Enlisted Personnel, Puget Sound Naval
        Shipyard, Bremerton, WA
To: Judge Advocate General

Subj: PETITION FOR A NEW TRIAL ICO LCDR WALTER F. FITZPATRICK,
        III, 551-90-4692/1110

Ref: (a) LCDR Fitzpatrick's ltr dtd 23 Feb 93
        (b) LCDR Fitzpatrick's ltr dtd 23 Sep 93

1. Request advise this command as to the status of LCDR
Fitzpatrick's request for a new trial contained in reference (a).

2. LCDR Fitzpatrick should be granted a new trial based on
documented evidence, dated as early as 23 November 1989, which
appears to have been withheld from him during and subsequent to
his trial and which could have been used by him in his defense or
in support of an earlier petition for a new trial.

3. As the Court-Martial Convening Authority for Puget Sound
Naval Shipyard I have reviewed documents related to LCDR
Fitzpatrick's Special Court-Martial. The documentation I have
seen is clear, compelling and suggests that command influence
inappropriately prejudiced the outcome of his trial. The fact
that the Preliminary Investigating Officer was also the adviser
to the Convening Authority and that two of the three court
members served on the Convening Authority's staff at the time of
trial substantiates this allegation and the allegations contained
in reference (b) and was clearly inappropriate in accordance with
the Manual for Courts-Martial.

4. In view of the fact that documentation which could have
altered the outcome of the trial has been withheld and is still
being withheld from LCDR Fitzpatrick, the two-year requirement
for submission of a petition for a new trial should be waived.

5. LCDR Fitzpatrick is a spectacular officer whose loss to the
Navy would be tragic, especially in light of the extraordinary
circumstances which surround the investigation of his alleged
offense and his trial. In the interest of justice and fair play
he should be granted a new trial. In the interest of the Navy we
should exhaust every effort conceivable to allow a truly
remarkable officer to continue his naval career. Time is of the
essence. A rapid response will be most appreciated.

D. R. PROULX

APPENDIX FOURTEEN

DEPARTMENT OF THE NAVY
NAVAL RESERVE CENTER
P. O. BOX 499
BREMERTON, WASHINGTON 98310-0121

IN REPLY REFER TO:

5800
Ser 00/339
28 Sep 94

From: Commanding Officer, Naval Reserve Center Bremerton, WA
To: Chief of Naval Personnel (Pers-253)

Subj: RETENTION ON ACTIVE DUTY PENDING DETERMINATION OF PETITION FOR A
      NEW TRIAL ICO LCDR WALTER F. FITZPATRICK III, 551-90-4692/1110

Ref: (a) CHNAVPERS WASHINGTON DC 211521Z SEP 94
      (b) CO Enlisted Personnel, Puget Sound Naval Shipyard, Bremerton
          WA ltr 5800 Ser 811/2447 of 02DEC93

1. Per reference (a), LCDR W. F. Fitzpatrick is scheduled to be separated
from active Naval service on 1 October 1994.

2. It is felt that LCDR Fitzpatrick should be granted a new trial based on
documented evidence dated as early as 23 November 1989, which appears to have
been withheld from him at the time and subsequent to his trial which could
have been used by him in his defense or in support of his earlier petition
for a new trial.

3. I have had the opportunity to review a significant amount of hard
documentary evidence relating to LCDR Fitzpatrick's Special Court Martial.
Based on the evidenciary material presented by LCDR Fitzpatrick it appears
that unlawful command influence is present and may have inappropriately
prejudiced the outcome of his trial. If this is the case, then the actions
of the Preliminary Investigating Officer and two of the three court members
who served on the Convening Authority's staff at the time of the trial were
in violation of Articles 6(c) and 37 of the UCMJ and the Manual for Court
Martial.

4. LCDR Fitzpatrick has faithfully and loyally served and supported the Navy
throughout the entire process leading to the determination to separate him on
1 October 1994. In the interest of justice and upon the review of the large
amount of unassailable documentary evidence, it is felt LCDR Fitzpatrick
should be authorized to remain on active duty pending the final outcome of the
submitted per DOD General Counsel's review/investigation and the response for
a new trial previously submitted per reference (b).

B. J. BODALY

309

APPRECIATIONS

⚓

WHAT CAN YOU SAY BEYOND the words "THANK YOU!"

My editor first! God bless you, Dominic McFarland Martin! Sergeant of Marines Timothy Joseph Harrington. Command Master Chief Poasa Fa'aita (*USS MARS* [AFS-1] and *USS Constellation* [CV- 64]). Sharon Rondeau, Post & Email owner and editor. Kit Langue for outstanding reporting. Paula Grubs, senior reporter to the *Cranberry Eagle* newspaper. James W. Cheevers, senior curator of the US Naval Academy Museum. Dr. Stephen Randolph (Col. US Air Force, Retired), historian of the US State Department.

BIBLIOGRAPHY

⚓

DR. WALTER FRANCIS FITZPATRICK JR.

Appleman, Roy E., Burns, James M., Gugeler, Russell A., and Stevens, John. (1995). *OKINAWA: The Last Battle*. New York: Barnes & Noble.

Astor, Gerald. (1995). *OPERATIONS ICEBERG: The Invasion and Conquest of Okinawa in World War II — An Oral History*. New York: Donald I. Fine, Inc.

Coale, Griffith Baily. (1942, Republished 1943 with Afterword). *North Atlantic Patrol: The Log of a Seagoing Artist*. New York: Farrar & Rinehart, Inc.

Diamond, Jon. (2017). *First Blood in North Africa: Operation Torch and the U.S. Campaign in Africa in World War II*: Guilford, Connecticut: Stockpole Books.

Helm, Thomas. (2001). *Ordeal by Sea: The Tragedy of the USS Indianapolis*. New York: Signet.

Nichols, Charles, Jr. and Shaw, Henry I. (1955). *Okinawa: Victory in the Pacific*.

Nashville, Tennessee, reprinted by The Battery Press, Inc.

O'Hara, Vincent P. (2015). *TORCH: North Africa and the Allied Path to Victory*. Annapolis, Maryland: Naval Institute Press.

Shaara, Jeffrey M. (2006). *The Rising Tide*. New York: Ballantine Books (a Random House Imprint).

Sledge, E.B. (1981 Republished 2007) *With the Old Breed: At Peleliu and Okinawa*. New York: Presidio Press (a division of Random House).

Yahara, Hiromichi. (1995). *The Battle for Okinawa: A Japanese Officer's Eyewitness Account of the Last Great Campaign of World War II*. New York: John Wiley & Sons, Inc.

USS INDIANAPOLIS

Caraley, Demetrios. (1966). *The Politics of Military Unification: A Study of Conflict and the Policy Process.* New York: Columbia University Press.

Dynes v. Hoover, 61 U.S. 20 How. 65 65 (1857).

Generous, William T. (1973). *Swords and Scales: The Development of the Uniform Code of Military Justice.* Port Washington New York: Kennikat Press.

Helm, Thomas. (1963). *Ordeal by Sea: The Tragedy of the U.S.S. Indianapolis.* New York: Dodd, Mead & Company.

Harrell, Edgar. *Out of the Depths. An Unforgettable WWII Story of Survival, Courage, and the Sinking of the USS Indianapolis.* Bloomington, MN, Behanay House.

Kurzman, Dan. (1990). *Fatal Voyage: The Sinking of USS Indianapolis.* New York: Pocket Books (Simon & Schuster).

Lech, Raymond B. (1982). *All the Drowned Sailors.* New York: Stein and Day.

Nelson, Peter. (2002) *Left For Dead: A Young Man's Search For Justice for the USS Indianapolis.* New York: Delacort Press (a division of Random House).

Newcomb, Richard F. (1958, Introduction and Afterword 2001) *Abandon Ship! The Saga of the U.S.S. Indianapolis, the Navy's Greatest Sea Disaster.* New York: Harper Collins.

Senate Armed Services Committee Report (September 14, 1999 | 106th Congress).

Smith, Holland M., and Finch, Percy. (1948). *Coral and Brass.* New York: Scribner.

Stanton, Doug. (2001). *In Harm's Way: The Sinking of the USS Indianapolis and the Extraordinary Story of Its Survivors.* New York: Henry Holt and Company, LLC.

U.S.S. Indianapolis FAMILIES (as compiled by Mrs. Mary Lou Murphy, wife of survivor Paul Murphy). (2008). *Lost At Sea But Not Forgotten: Memories of the Men Who Gave Their Lives In the Sinking of the USS Indianapolis (CA-35).* Indianapolis, Indiana: Printing Partners, Inc.

Vincent, Lynn, & Vladic, Sara. (2018). *Indianapolis: The True Story of the Worst Sea Disaster in U.S. Naval History and the Fifty-Year Fight to Exonerate An Innocent Man.* New York: Simon & Schuster.

BOOKS (NONFICTION)

Adkins, Lesley, & Roy A. Adkins. (1994). *Handbook to Life in Ancient Rome.* Oxford: Oxford University Press.

Allen, Robert, L. (1993). *The Port Chicago Mutiny.* New York: Amistad.

Allen, Thomas B. & Dickson, Paul. (2004). *The Bonus Army: An American Epic.* New York: Warner and Company.

Alexander, Caroline. (2003). *The Bounty: The True Story of the Mutiny on the Bounty.* New York: Viking.

Ambrose, Stephen E. (1997). *Citizen Soldiers: The U.S. Army from the Normandy Beaches to the Bulge to the Surrender of Germany.* New York: Simon & Schuster.

Ambrose, Stephen E. (1996). *Undaunted courage: Meriwether Lewis, Thomas Jefferson, and the opening of the American West.* New York: Simon and Schuster.

Andersonville Diary & Memoirs of Charles Hopkins, The. (1988). Kearney, NJ: Belle Grove Publishing Co.

Barnett, Louise. (2000). *Ungentlemanly Acts: The Army's Notorious Incest Trial.* New York: Hill and Wang.

Barone, Michael. (2007). *Our First Revolution: The Remarkable British Upheaval That Inspired America's Founding Fathers.* New York: Three Rivers Press.

Bilton, Michael & Sim, Kevin. (1992). *Four Hours at My Lai.* New York: Viking Press.

Bilton, Michael & Sim, Kevin. *Black's Law Dictionary* (Seventh Ed.). (1991). St. Paul, MN: West Group.

Blackstone, William. (1765). *Commentaries on the Laws of England* (Vol. 1). Chicago: University of Chicago Press.

Bluejacket's Manual, The (13th ed.). (1946). Annapolis, MD: Naval Institute Press. (First published in 1902.)

Bluejacket's Manual, The (18th ed.). (1968). Annapolis, MD: Naval Institute Press.

Bluejacket's Manual, The (19th ed.). (1973). Annapolis, MD: Naval Institute Press.

Bluejacket's Manual, The (21st ed.). (1990). Annapolis, MD: Naval Institute Press.

Blumenson, Martin. (1985). *Patton: The Man Behind the Legend 1885–1945*. New York: William Morrow.

Boyington, Gregory, "Pappy." (1958). *Baa Baa Black Sheep*. New York: Bantam.

Bradley, James. (2003). *Flyboys: A True Story of Courage*. New York: Little, Brown.

Bracton, Henry. Brown, Don. (2019). *Travesty of Justice: The Shocking Prosecution of LT. Clint Lorance*. Denver, Colorado: Wild Blue Press.

Buckler, John & Hill, Bennett D. & McKay, John P. (2003). *A History of Western Society Since 1300* (7th ed.). New York: Houghton Mifflin.

Buell, Thomas, B. (1980). *Master of Sea Power: A Biography of Fleet Admiral Ernest J. King*. New York: Little, Brown and Company.

Burstein, Andrew. (2003). *The Passions of Andrew Jackson*. New York: Vintage Books.

Butler, Smedley Darlington. (1935). *War Is A Racket*.

Byrne, Edward M. (1970). *Military Law* (3rd ed.). Annapolis MD: Naval Institute Press.

Byrne, Edward M. (1981). *Military Law* (3rd ed.). Annapolis, MD: Naval Institute Press.

Calhoon, Robert McCluer. (1973). *The Loyalists In Revolutionary America, 1760–1781*. New York: Harcourt Brace Jovanovich.

Calhoun, Raymond, C. (1993). *Tin Can Sailor: Life Aboard the USS Sterett 1939–1945*. Annapolis, MD: Naval Institute Press.

Calhoun, Raymond C. (1981). *Typhoon: The Other Enemy: The Third Fleet and the Pacific Storm of December 1944*. Annapolis, MD: Naval Institute Press.

Cawthorne, Nigel. (2009). *Kings & Queens of England: From the Saxon Kings to the House of Windsor*. New York: Metro Books.

Chambers II, John Whiteclay. (Ed.) (1999). *The Oxford Companion To American Military History*. New York: Oxford University Press.

Childs, John. (1980). *The Army, James II and the Glorious Revolution*. Manchester, England: Manchester University Press.

Churchill, Winston Spencer. (1956). *A History of the English-Speaking Peoples: The Birth of Britain*. New York: Dodd, Mead & Company.

Churchill, Winston S. (1957). *Churchill's History of the English-Speaking Peoples*. New York: Barnes and Noble Books.

Clancy, Tom & Koltz & Stiner, Carl. (2002). *Shadow Warriors: Inside the Special Forces*. New York: G.P. Putnam's Sons.

Clausewitz, Carl von. (1997). *On War* (J. J. Graham, Trans.). England, Hertfordshire: Wordsworth Classics.

Cohen, Eliot, A. (2002). *Supreme Command: Soldiers, Statesmen, and Leadership in Wartime*. New York: The Free Press.

Collier, Christopher & James Lincoln. (1986). *Decision In Philadelphia: The Constitutional Convention of 1787*. New York: Ballantine Books.

Connelly, Owen. (2002). *On War and Leadership: The Words of Combat Commanders from Frederick the Great to Norman Schwarzkopf*. Princeton, NJ: Princeton University Press.

Connery, Horace J. & Laudenslager, John M. & Mann, Gregory J. & McCandless, Bruce & Mulholland, Frank J. & Wolfe, Malcolm E. (1959). *Naval Leadership* (2nd ed.). Annapolis, MD: Naval Institute Press.

Cope, Harley. (1951). *The Naval Officer's Manual*. (2nd ed.). Harrisburg, PA: The Military Service Publishing Company.

Cristol, A. Jay: (2002). *The Liberty Incident: The 1967 Israeli Attack on the U.S. Navy Spy Ship*. Washington, D.C.: Brassey's, Inc.

Davidson, Michael J. (1999). *A guide to military criminal law.* Annapolis, Maryland: Naval Institute Press.

Davis, William C. (1999). *Lincoln's Men: How President Lincoln Became Father to an Army and a Nation.* New York: Touchstone.

Davis, Burke. (1967). *The Billy Mitchell Affair.* New York: Random House.

Decter, Midge. (2003). *Rumsfeld.* New York: Harper Collins.

Diggins, Patrick John. (2003). *John Adams.* (Arthur M. Schlesinger, Jr.) (Ed.). New York: Times Books.

Di Mona, Joseph. (1972). *Great Court-Martial Cases.* New York: Grosset and Dunlap.

Eisenhower, Dwight D. (1967). *At Ease: Stories I Tell to Friends.* New York: Avon Books.

Edmondson, J. R. (2000). *The Alamo Story: From Early History to Current Conflicts.* The Rowman & Littefield Publishing Group, Inc.: Republic of Texas Press.

Ensign, Tod. (2004). *America's Military Today: The Challenge of Militarism.* New York: The New Press.

Everitt, Anthony. (2001). *Cicero: The Life and Times of Rome's Greatest Politician.* New York: Random House.

Farah, Joseph. (2003). *Taking America Back: A Radical Plan to Revive Freedom, Morality, and Justice.* Nashville, TN: Thomas Nelson.

Farber, Daniel A. (2003). *Lincoln's Constitution.* Chicago, IL: The University of Chicago Press.

Ferling, John. (1992). *John Adams: A Life.* New York: Henry Holt and Co.

Finn, James. (Ed.). (1971). *Conscience and Command.* New York: Random House.

Flexner, James Thomas. (1953). *The Traitor and the Spy.* Boston: Little, Brown, & Co.

Foote, Shelby. (1986). *The Civil War: A Narrative.* New York: Vintage Books.

Ford, Corey. (1965). *A Peculiar Service.* Boston: Little, Brown, & Co.

Galvani, William. (1999). *Mainsail to the Wind: A Book of Sailing Quotations*. Dobbs Ferry, NY: Sheridan House, Inc.

Generous, William T. Jr. (1973). *Swords and Scales: The Development of the Uniform Code of Military Justice*. Port Washington, New York: National University Publications.

Gibbon, Edward. (2000). *The History of the Decline and Fall of the Roman Empire: Abridged Edition*. New York: Penguin. (Full edition first published in 1776.)

Gillingham, John. (2005). *The Wars of the Roses: Peace & Conflict in 15th Century England*. London: Phoenix Press.

Goldstein, Joseph & Marshall, Burkr & Schwartz, Jack. (1976). *The My Lai Massacre and Its Cover-Up*. New York: Free Press.

Green, Vincent. (1995). *Extreme Justice*. New York: Pocket Books.

Greene, Joshua M. (2003). *Justice at Dachau: The Trials of an American Prosecutor*. New York: Broadway Books.

Grossman, Dave. (1995). *On Killing: The Psychological Cost of Learning to Kill in War and Society*. New York: Back Bay Books.

Gutmann, Stephanie. (2000). *The Kinder, Gentler Military: Can America's Gender-Neutral Fighting Force Still Win Wars?* New York: Scribner.

Guttridge, Leonard F. (1992). *Mutiny: A History of Naval Insurrection*. Annapolis, MD: Naval Institute Press.

Hackworth, David H. & Sherman, Julie. (1989). *About Face*. New York: Simon and Schuster.

Hackworth, David M. & Tom Mathews. (1996). *Hazardous Duty*. New York: Avon Books.

Hale, Matthew. (1739, republished 1971). *The History of the Common Law of England*. Chicago, Illinois: University of Chicago Press, p. 4. Also retrievable from: https://constitution.org/1-Constitution/cmt/hale/history_common_law.htm.

Hammer, Richard. (1971). *The Court-Martial of Lt. Calley*. New York: Coward, McCann & Geoghegan.

Hansen, Harry. (1970). *The Boston Massacre: An Episode of Dissent and Violence*. New York: Hastings House.

Hart, Gary. (1998). *The Minuteman: Restoring an Army of the People*. New York: The Free Press.

Hartle, Anthony E. (2004). *Moral Issues in Military Decision Making*. Lawrence, KS: University Press of Kansas.

Hatch, Robert McConnell & Higginbotham, Don. (1986). *Major John Andre: A Gallant in Spy's Clothing*. Boston: Houghton Mifflin.

Hayford, Harrison. (1959). *The Somers Mutiny Affair*. Englewood Cliffs, NJ: Prentice Hall.

Headley, Lake, & Hoffman, William. (1989). *The Court-Martial of Clayton Lonetree*. New York: Henry Holt.

Heemstra, Thomas, S. (2002). *Anthrax: A Deadly Shot in the Dark*. Lexington, KY: Crystal Communications.

Helm, Thomas. (2001). *Ordeal by Sea: The Tragedy of the USS Indianapolis*. New York: Signet.

Hersh, Seymour M. (2004). *Chain of Command*. New York: Harper Collins.

Hinckley, Barbara. (1985). *Problems of the Presidency*. Glenview, IL: Scott, Foresman and Company.

Hoffman, Robert L. (1980). *More Than a Trial: The Struggle Over Captain Dreyfus*. New York: The Free Press (A division of Macmillan Publishing Co.).

Hoffmann, George F. (1993). *Cold War Casualty: The Court-Martial of Major General Robert W. Grow*. Kent, OH: Kent State University Press.

Holland, Tom. (2003). *Rubicon: The Last Years of the Roman Republic*. New York: Doubleday.

Holmes, Oliver Wendell. (1963). *The Common Law*. New York: Little, Brown and Company.

Homan, Gerlof D. (1994). *American Mennonites and the Great War 1914–1918*. Scottdale, PA: Herald Press.

Huchthausen, Peter A. (2003). *America's Splendid Little Wars: A Short History of U.S. Military Engagements: 1975–2000*. New York: Viking.

Huie, William Bradford (1954). *The Execution of Private Slovik*. Yardley, Pennsylvania: Westholme.

Huntington, Samuel P. (1957, 2002). *The Soldier and the State: The Theory and Politics of the Civil-Military Relations* (19th ed.). New York: Random House.

Huntington. (2004). *Who Are We? : America's National Identity and the Challenges It Faces*. New York: Simon and Schuster.

Iglesias, David & Seay, Darin (2008). *In Justice: Inside the Scandal that Rocked the Bush Administration*. Hoboken, New Jersey: John Wiley & Sons, Inc.

Jackson, Percival, E. (Ed.). (1962). *The Wisdom of the Supreme Court*. Norman, OK: University of Oklahoma Press.

Jager, Eric. (2004). *The Last Duel: A True Story of Crime, Scandal, and Trial By Combat in Medieval France*. New York: Broadway Books.

Janowitz, Morris. (1971). *The Professional Soldier: A Social and Political Portrait*. New York: The Free Press.

Johnson, Chalmers A. (2004). *The Sorrows of Empire: Militarism, Secrecy, and the End of the Republic*. New York. Metropolitan Books.

Johnson, Paul. (1997). *A History of the American People*. New York: Harper Perennial.

Jones, Dan. (2014). *The Wars of the Roses: The Fall of the Plantagenets and the Rise of the Tudors*. New York: Penguin Books.

Kagan, Donald. (2003). *The Peloponnesian War*. New York: Penguin Books.

Kaplan, Robert D. (2002). *Warrior Politics: Why Leadership Demands a Pagan Ethos*. New York: Vintage Books.

Karnow, Stanley. (1983). *Vietnam: A History*. New York: The Viking Press.

Kastenberg, Joshua, E. (2009). *The Blackstone of Military Law: Colonel William Winthrop*. Lanham, Maryland: The Scarecrow Press, Inc., p. 137.

Keegan, John. (1976). *The Face of Battle*. New York: Penguin Books.

Kelly, C. Brian. (1999). *Best Little Stories from the American Revolution*. Nashville, TN: Cumberland House.

Kelly, C. Brian. (1994). *Best Little Stories from the Civil War*. Nashville, TN: Cumberland House.

Ketchum, Richard M. (1973). *Will Rogers: His Life and Times*. New York: Simon and Schuster.

King, Dean & Estes, J. Worth & Hattendorf, John B. (2000). *A Sea of Words: A Lexicon and Companion to the Complete Seafaring Tales of Patrick O'Brian* (3rd ed.). New York: Henry Holt and Company.

Kirschke, James J. (2005). *Gouverneur Morris: Author, Statesman, and Man of the World*. New York: St. Martin's Press.

Knappman, Edward W. (Ed.). (1994). *Great American Trials*. New York: Barnes and Noble.

Kohn, Richard H. (1975). *Eagle and Sword: The Beginning of the Military Establishment in America*. New York: Free Press.

Lardner, Rex. (1966). *Ten Heroes of the Twenties*. New York: G.P. Putman's Son's.

Layton, Edwin T. (1985). *"And I Was There": Pearl Harbor and Midway – Breaking the Secrets*. New York: William Morrow.

Leckie, Robert. (1992). *George Washington's War: The Saga of the American Revolution*. New York: Harper Perenial.

Levin, Mark. (2005) *Men in Black: How the Supreme Court Is Destroying America*. Washington, D.C.: Regnery.

Leyva, Meredith. (2003). *Married to the Military: A Survival Guide for Military Wives, Girlfriends, and Women In Uniform*. New York: Simon and Schuster.

Lipsky, David. (2003). *Absolutely American: Four Years at West Point*. New York: Houghton Mifflin.

Livy. (1960). *The Early History of Rome: Books I – V of the History of Rome from Its Foundations*. (Aubrey de Sélincourt, Trans.). New York: Penguin Books.

Lord, Francis, A. (1960). *They Fought for the Union*. Westport, CT: Greenwood Press.

Lovell, John P. & Kronenberg, Philip S. (Eds.) (1974). *New Civil-Military Relations: The Agonies of Adjustment to Post-Vietnam Realities*. New Brunswick, NJ: Transaction Books.

Lovette, Leland P. (1939). *Naval Customs, Traditions, and Usage* (3rd ed.). Annapolis, MD: Naval Institute Press.

Lovette, Lelans P. (1959). *Naval Customs, Traditions, and Usage* (4th ed.). Annapolis, MD: Naval Institute Press.

Lowry, Thomas P. (1999). *Don't Shoot That Boy! Abraham Lincoln and Military Justice*. Mason City, IA: Savas.

Lowry, Thomas P. (1997). *Tarnished Eagles: The Courts-Martial of Fifty Union Colonels and Lieutenant Colonels*. Mechanicsburg, PA: Stackpole Books.

Lurie, Jonathan (2001). *Military Justice in America: The U.S. Court of Appeals for the Armed Forces, 1775–1980*. Lawrence: University Press of Kansas.

Manchester, William. (1978). *American Caesar: Douglas MacArthur 1880–1964*. New York: Random House.

Manaster, Kenneth A. (2001). *Illinois justice: The Scandal of 1969 and the Rise of John Paul Stevens*. Chicago: University of Chicago Press.

Matsumoto, Gary. (2004). *Vaccine A: The Covert Government Experiment that's Killing Our Soldiers*. New York: Basic Books.

Marshall, Samuel Lyman Atwood. (1947). *Men Against Fire: The Problem of Battle Command*. Norman, OK: University of Oklahoma Press.

McCain, John. (1999). *Faith of My Fathers*. New York: Perennial.

McCullough, David. (2001). *John Adams*. New York: Simon and Schuster.

McCullough, David. (1992). *Truman*. New York: Simon and Schuster.

McFarland, Philip James. (1985). *Sea Dangers: The Affair of the Somers*. New York: Schocken Books.

McMaster, H.R. (1997). *Dereliction of Duty: Lyndon Johnson, Robert McNamara, the Joint Chiefs of Staff, and the Lies That Led to Vietnam*. New York: Harper Perennial.

McMichael, William H. (1997). *The Mother of All Hooks: The Story of the U.S. Navy's Tailhook Scandal.* New Brunswick, NJ: Transaction Publishers.

Melton, Buckner F. (2003). *A Hanging Offense: The Strange Affair of the Warship Somers.* New York: Free Press.

Melville, Herman. (1963). *White-Jacket; or, The World in a Man-of-War.* (Harrison Hayford & Hershel Parker & G. Thomas Tanselle, Eds.). Evanston, IL: Northwestern University Press. (First published in the United States in 1850.)

Middlekauff, Robert. (1982). *The Glorious Cause, the American Revolution, 1763–1789.* New York: Oxford University Press.

Military Quotation Book, The. (1990). Charlton, James (Ed.). New York: St. Martin's Press.

Morris, Richard B. (1987). *The Forging of the Union 1781–1789.* New York: Harper & Row.

Napolitano, Andrew, P. (2004). *Constitutional Chaos: What Happens When the Government Breaks Its Own Laws.* Nashville, Tennessee: Nelson Current.

Nimitz, Chester, W. & Potter, E. B. (Eds.). (1960). *Sea Power: A Naval History.* Englewood Cliff, NJ: Prentice-Hall, Inc.

O'Leary, Jeff. (2001). *Taking the High Ground: Military Moments with God.* Colorado Springs, Colorado: Victor.

Osiel, Mark (1999). *Obeying Orders: Atrocity, Military Discipline, and the Law of War.* New Brunswick, NJ: Transaction Publications.

Owens, Bill & Offley, Ed. (2000). *Lifting the Fog of War.* New York: Farrar, Straus and Giroux.

Patterson, Robert "Buzz." (2003). *Dereliction of Duty: The Eyewitness Account of How Bill Clinton Compromised America's National Security.* Washington, D.C.: Regnery.

Patterson, Robert "Buzz." (2004). *Reckless Disregard: How Liberal Democrats Undercut Our Military, Endanger Our Soldiers, and Jeopardize Our Security.* Washington, D.C.: Regnery.

Paine, Thomas. (1737–1809). *Rights of Man, Common Sense, and Other Writings.* (Mark Philp, Ed.). New York: Oxford University Press.

Parrish, Robert D. & Andreacchio, N.A. (1991). *Schwarzkopf: An Insider's View of the Commander and His Victory.* New York: Bantam Books.

Perret, Geoffrey. (2004). *Lincoln's War: The Untold Story of America's Greatest President as Commander In Chief.* New York: Random House.

Persico, Joseph, E. (1994). *Nuremberg: Infamy on Trial.* New York: Penguin Books.

Philbrick, Nathaniel (2003). *Sea of Glory: America's Voyage of Discovery: The U.S. Exploring Expedition 1838–1842.* New York: Penguin Books.

Polmar, Norman & Allen, Thomas B. (1991). *World War II: Americans at War, 1941–1945.* New York: Random House.

Potter, E.B. (1985). *Bull Halsey.* Annapolis, MD: Naval Institute Press.

Potter, E.B. (1976). *Nimitz.* Annapolis, MD: Naval Institute Press.

Priest, Dana. (2003). *The Mission: Waging War and Keeping Peace with America's Military.* New York: W.W. Norton.

Rakove, Jack N. (1996). *Original Meanings: Politics and Ideas in the Making of the Constitution.* New York: Vintage Books.

Randall, Willard Sterne. (1990). *Benedict Arnold: Patriot and Traitor.* New York: Dorset.

Ransom, John L. (9186). *John Ransom's Andersonville Diary.* Middlebury, VT: Paul S. Erickson.

Rampersad, Arnold. (1997). *Jackie Robinson: A Biography.* New York: Alfred A. Knopf.

Rehnquist, William H. (1998). *All the Laws but One: Civil Liberties in Wartime.* New York: Alfred A. Knopf.

Rehnquist, William H. (2001). *The Supreme Court* (Rev. ed.). New York: Alfred A. Knopf.

Roberts, Paul Craig & Stratton, Lawrence M. (2000). *The Tyranny of Good Intentions: How Prosecutors and Bureaucrats are Trampling the Constitution In the Name of Justice.* Roseville, CA: Forum.

Robinson, Greg. (2001). *By Order of the President: FDR and the Internment Of Japanese Americans*. Cambridge, MA: Harvard University Press.

Rodgers, Ledyard William. (1940). *Naval Warfare Under Oars, 4th to 16thCenturies: A Study of Strategy, Tactics, and Ship Design*. Annapolis, Maryland: Naval Institute Press.

Rooney, David. (1999). *Military Mavericks: Extraordinary Men of Battle*. London, England: Cassell and Co.

Santosuosso, Antonio. (2004). *Barbarians, Marauders, and Infidels: The Ways of Medieval Warfare*. Boulder, CO: Westview Press.

Scott, Richard. (1987). *Jackie Robinson*. (Nathan Irvin Huggins, Ed.). New York: Chelsea House.

Schaeffer, Frank & John. (2002). *Keeping Faith: A Father-Son Story About Love and the United States Marine Corps*. New York: Carroll and Graf.

Schecter, Barnet. (2002). *The Battle For New York: The City at the Heart of the American Revolution*. New York: Walker and Company.

Sheehan, Niel. (1971). *The Arnheiter Affair*. New York: Random House.

Sherrill, Robert. (1969, 1970). *Military Justice Is to Justice as Military Music Is to Music*. New York: Harper & Row.

Shilts, Randy (1993). *Conduct Unbecoming: Gays and Lesbians in the U.S. Military*. New York: St. Martin's.

Sloan, Elinor, C. (2002). *The Revolution in Military Affairs: Implications for Canada and NATO*. Montreal, Canada: McGill-Queen's University Press.

Snedeker, James. (1954). *A Brief History of Courts-Martial*. Annapolis, Maryland: Naval Institute Press.

Tacitus. (2003). *The Annals and The Histories*. (Moses Hadas, Ed.). (Alfred John Church & William Jackson Brodribb, Trans.). New York: The Modern Library.

Thomas, Evan. (2003). *John Paul Jones: Sailor, Hero, Father of the American Navy*. New York: Simon and Schuster.

Timberg, Robert. (1995). *The Nightingale's Song*. New York: Touchstone.

Toland, John. (1982). *Infamy: Pearl Harbor and Its Aftermath*. New York: Berkley Books.

Tolchin, Martin & Susan. (1992). *Selling Our Security*. New York: Alfred A. Knopf.

Tsouras, Peter, G. (2000). *The Greenhill Dictionary of Military Quotations*. Mechanicsburg, PA: Stackpole Books.

Turner, Thomas Reed. (1982). *Beware the People Weeping: Public Opinion and the Assassination of Abraham Lincoln*. Baton Rouge: Louisiana State University Press.

Valle, James E. (1980). *Rocks & Shoals: Naval Discipline in the Age of Fighting Sail*. Annapolis, Maryland: Naval Institute Press.

Van de Water, Frederic F. (1954). *The Captain Called It Mutiny*. New York: Washburn.

Van Doren, Carl. (1973). *The Secret History of the American Revolution*. New York: Augustus M. Kelley.

Vistica, Gregory. (1997). *Fall From Glory: The Men Who Sank the U.S. Navy* (Rev. ed.). New York: Touchstone Books.

Waddle, Scott, & Abraham, Ken. (2002). *The Right Thing*. Brentwood, TN: Integrity.

Waller, Douglas (2004). *A Question of Loyalty: Gen. Billy Mitchell and the Court-Martial that Gripped the Nation*. New York: Harpers Collins.

Walton, Clifford. (1894). *History of the British Standing Army. A.D. 1660 to 1700*. London: Harrison and Sons.

West, Luther C. (1977). *They Call It Justice: Command Influence and the Court-Martial System*. New York: Viking.

Webb, James (2004). *Born Fighting: How the Scots-Irish Shaped America*. New York: Broadway Books.

Webster, Mary E. (Ed.). (1999). *The Federalist Papers in Modern Language*. Bellevue, WA: Merrill Press.

Wightman, William. (1969). *Minutes of a Conspiracy Against the Liberties of America*. New York: Arno Press.

Wills, Garry (1999). *A Necessary Evil: A History of American Distrust of Government*. New York: Simon and Schuster.

Wills, Garry. (1992). *Lincoln at Gettysburg: The Words that Remade America*. New York: Touchstone.

Winthrop, William. (1920). *Military Law and Precedents* (2nd ed.) (Vol. II). Washington, D.C.: Beard Books.

Woodward, Bob & Armstrong, Scott. (1979). *The Brethren: Inside the Supreme Court*. New York: Simon and Schuster.

Woodward, Bob & Armstrong, Scott. (1991). *The Commanders*. New York: Simon and Schuster.

Yarmolinsky, Adam. (1971). *The Military Establishment: Its Impacts on American Society*. New York: Harper Colophon Books.

Zobel, Hiller B. (1970). *The Boston Massacre*. New York: Norton & Company.

Zola, Emile. (1996). *The Dreyfus Affair: "J'accuse" and other writings*. (Alain Paigès, Ed.). (Eleanor Levieux, Trans.). New Haven, CT: Yale University Press.

BOOKS (FICTION)

Arvin, Nick. (2005). *Articles of War*. New York: Doubleday.

Cobb, Humphrey (1987). *Paths of Glory*. Athens, Georgia: University of Georgia Press.

Conroy, Pat. (1980). *The Lords of Discipline*. New York: Houghton Mifflin.

Heinlein, Robert A. (1959). *Starship Trooper*. New York: G. P. Putnam's Sons.

Heller, Joseph. (1955). *Catch-22*. New York: Simon and Schuster.

Jones, James. (1951). *From Here to Eternity*. New York: Dell Publishing.

Kipling, Rudyard. (1892, 1893, 1899, 1917). *Departmental Ditties and Ballads and Barrack-Room Ballads*. New York: McMillan And Co, "PRELUDE."

McCullough, Colleen. (1990). *The First Man in Rome*. New York: William Morrow and Co., Inc.

McCullougn, Coleen. (1991). *The Grass Crown*. New York: William Morrow and Co., Inc.

Mailer, Norman. (1948). *The Naked and the Dead*. New York: Henry Holt and Co.

Melville, Herman (1986). *Billy Budd and Other Stories*. New York: Penguin Books.

Myrer, Anton. (1968). *Once an Eagle*. New York: Holt, Rinehart, and Winston.

Ponicsan, Darryl. (1970). *The Last Detail*. New York: Signet.

Pressfield, Steven. (1999). *Gates of Fire: An Epic Novel of the Battle of Thermopylae*. New York: Bantam Books.

Pressfield, Stever. (2000). *Tides of War: A Novel of Alcibades and the Peloponnesian War*. New York: Bantam Books.

Shaara, Jeff. (2001). *Rise to Rebellion*. New York: Ballantine Books.

Wouk, Herman. (1951). *The Caine Mutiny*. New York: Doubleday and Company, Inc.

Wouk, Herman. (1954). *The Caine Mutiny Court-Martial*. New York: Doubleday & Co., Inc.

PERIODICALS

Adde, Nick. (1997, 27 January). "'Undue Command Influence' Issue Still Alive and Well." *Army Times*, p. 11.

Ambrose, Stephen E. (1999, August 9). "The End of the Draft, and More: Common Identity Is Lost." *National Review*, 35–36.

Auster, Bruce B. & Cary, Peter. (1992, July 13). "What's Wrong with the Navy: Beset By Scandals, the Service Faces an Uncertain Future." *U.S. News & World Report*, 22–29.

Bacevich, Andrew, J. "Losing Private Ryan: Why the Citizen-Soldier is MIA." *National Review*, 32–34.

Barnes, Julian E. & Mazzettti, Mark & Pound, Edward T. (2004, May 24). "Inside the Iraq Prison Scandal." *U.S. News & World Report*, 19–20, 22, 24–28.

Brown, Richard, C. (1975, August). "Three Forgotten Heroes." *American Heritage*, p. 25.

Cannon, Angie & Ragavan, Chitra. (2004, May 24). "A Big Legal Mess, Too." *U.S. News & World Report*, 29.

Cary, Peter. (1992, July 13). "Double Cross: How Not to Do Business." *U.S. News & World Report*, 29, 31.

Cary, Peter. (1992, November 9). "Navy Justice: Why the Service Can't Police Itself." *U.S. News & World Report*, 46–48, 58–60, 63.

Cory, Peter. (1992, November 9). "The China Lake Affair." *U.S. News & World Report*, 63–65.

Charen, Mona. (2000, April 10). "General's Harassment Charge is a Clinto Legacy." *The Tacoma News Tribune*, p. B4.

"Courts-Martial of Black Sailors in 1944 to Stand." (1994, January 17). *Navy Times*, p. 18.

Derbyshire, John. (2001, April 16). "Is This All We Can Be?" *National Review*, 30–31, 34–35.

Duffy, Michael & Thompson, Mark. (2003, January 5). "Warrior in Chief." *Time*, 82–85, 87–88, 91–92, 94–96.

Duncan, Robert F. & Fiscus, Thomas J. & Haynes, William J. & Lohr, Michael F. & Sandkuhler, Kevin M. (2003). "Unequal Justice [Letter to the editor]." *U.S. News & World Report*, p. 9.

Fenwick, Ben C. (1987, February). "The Plot to Kill Washington." *American History Illustrated*, 8–12.

Fleming, Thomas J. (1969, December). "The Boston Massacre." *American Heritage*, 6–11, 102–111.

Gibbs, Nancy. (2003, January 5). "Person of the Year: The American Soldier." *Time*, 32–36, 41.

Hillman, Beth (2002 May/June). "Chains of Command." *Legal Affairs*, Vol. 1, No. 1, 50–52.

Kimmelman, Benedict B. (1987, September/October). "The Example of Private Slovik." *American Heritage*, 97–104.

Maier, Timothy W. (2001, May 14). "Anthrax Vaccine: Ducking the Shot." *Insight*, 20–21.

Murphy, Mary. (2004, January 10). "Murders, Mysteries, and the Military." *TV Guide*, 44–48.

Nachtwey, James. (2003, January 5). "Photo Essay: Snapshots of a Soldier's Life." *Time*, 42–57.

Nollinger, Mark. (2001, March 17). "Disaster at Sea." *TV Guide*, 49–50, 52.

O'Meara, Kelly Pat. (2001, May 14). "How Just Is Our Military Justice?" *Insight* 22–23, 39.

Philpott, Tom. (1979). "CMA Chief, Services in Escalating Feud." *Navy Times*, p. 15.

Pound, Edward, T. (2002, December 16). "Unequal Justice: Military Courts Are Stacked to Convict – Yet Brass Get Off Easy." *U.S. News & World Report*, 18–22, 24–30.

Ratnesar, Romesh & Weisskopf, Michael. (2003, January 5). "One Platoon's Story: On Alert." *Time*, 58–66, 68, 71, 76, 78, 80–81.

Schwed, Mark. (2003, May 3). "JAG Flies High." *TV Guide*, 26–29.

Szegedy-Maszak, Marianne. (2004, May 24). "Sources of Sadism." *U.S. News & World Report*, 30.

Timmerman, Kenneth R. (2001, May 14). "Secrets of the Strategic Review." *Insight*, 18–19.

Weiner, Ed. (2004, January 10). "TV Loves a Man (or a Woman) in Uniform." *TV Guide*, 36–38, 40.

NEWSPAPERS | MAGAZINES

Gage, Nicholas. (January 2007). "Race Against Terror." *Vanity Fair*, p. 64.

Littell, Robert. (August 1970). "Military Justice on Trial." *Newsweek*, p. 18.

Pound, Edward T. (6 December 2002). "Unequal Justice: Why America's Military Courts Are Stacked to Convict." *U.S. News*, p. 18

Rudolph, Ileana. (December 8, 2008). "The Real Criminal Minds: The Experts Keep Procedures on the Right Law." p. 35.

ACADEMIC PUBLICATIONS | JOURNALS

Ansel, Samuel T. (1919). "Military Justice." *The Cornell Law Quarterly*, Volume V (Number 1), pp. 1 - 17.

Hagan, William R. (1986). "The Yet-Unpaid Debt of King Gustavus Adolphus: The Development of Military Law in Europe During the Cinquecento." Retrieved from: http://www.pegc.us/_LAW_/Hagan_Military_Law.pdf. (Colonel Hagan's paper is an update to his 1986 paper titled "Overlooked Textbooks Jettison Some Durable Military Law Legends." *Military Law Review*, Vol 113.)

Hoyler, H. M. (June 1961). "The Development of Military Law – Your Job and Mine." *U.S. Naval Proceedings*, June 1961, p. 32.

Sullivan, Dwight, H. (1998). "Playing the Numbers: Court-Martial Panel Size and the Military Death Penalty." *Military Law Review*, Volume 158, pp. 1–47.

"The Armed Forces: Who They Are." (2003, January 5). *Time*, 98–99.

INTERNET

Gilbert, Michael H. (1997–1998). "The Military and the Federal Judiciary: An Unexplored Part of the Civil-Military Relations Triangle." Retrieved October 30, 2001. **http://www.usafa.af.mil/dfl/documents/milfed.doc.**

Gonzales, Alberto R. (2001, November 30). "Martial Justice, Full and Fair." *New York Times*. Retrieved December 2, 2001. **http://www.nytimes.com/2001/11/30/opinion/30GONZ.html**

Jerrell, Kit. "Former JAG Attorney Implicates JAG Corps." Retrieved December 31, 2006. http://www.leatherneck.com/forums/showthread/php?39351

GOVERNMENT PUBLICATIONS

Department of the Army Headquarters 75th Field Artillery Brigade. (2001, August 13). "Appointment as Army Regulation (AR) 15-6 Investigating Officer."

Department of Defense. (1988). "The Armed Forces Officer.

Manual for Courts-Martial." United States, Washington, D.C.: GPO, 1951.

The United States Constitution. (1787).